The Mathematical Enigma of Guamán Poma's Yupana

Cinzia Florio

The Mathematical Enigma of Guamán Poma's Yupana

A Hundred Years' Riddle

Springer

Cinzia Florio
Moliterno (PZ), Italy

ISBN 978-3-031-92661-7 ISBN 978-3-031-92662-4 (eBook)
https://doi.org/10.1007/978-3-031-92662-4

Cover illustration: Reworked image of page 362 from "El primer Nueva Corónica y Buen Gobierno by Felipe Guamán Poma de Ayala".

This Springer imprint is published by the registered company Springer Nature Switzerland AG
The registered company address is: Gewerbestrasse 11, 6330 Cham, Switzerland

Ai discendenti degli antichi Incas
alla ricerca delle
radici perdute

Foreword

Like most researchers investigating numbers, I apply the theories and methods of my discipline to my field of study. As my discipline is cognitive archaeology, I focus on the material devices used to represent and manipulate numbers—fingers, tallies, tokens, and numerical notations. This approach differs from mainstream research into numerical cognition, which is dominated by cognitive psychologists looking at how the brain functions in numbers, developmental psychologists examining at how children acquire numbers, comparative psychologists investigating what non-human species have in terms of numbers, and linguists looking at how language works in numbers. Pervading the mainstream is the assumption that numbers are produced by the brain. In this neurocentric construct, the material devices used for numbers are mere reflections of what brains think up and later externalize to them; the devices are simply passive recipients of whatever brains conceive.

The problem with the "brain-only" approach is that it cannot satisfactorily explain cross-cultural variability. Numbers vary between societies, both in how they are organized and whether there even are any to be organized. Some societies, instead of using decimal organization, use sexagesimal (base-60), vigesimal (base-20), duodecimal (base-12), octal (base-8), senary (base-6), quaternary (base-4), or binary (base-2). Some societies, instead of counting to very high numbers or infinity, count no higher than "two" or "three," and one society in the Amazon is said not to have any numbers at all (Everett 2005). Here brain-focused explanations have proven particularly inadequate, for brains in societies with numbers are seen as somehow more capable, to the disadvantage of brains in societies without numbers. Nor have brain-focused investigations been able to explain why numbers are organized decimally in some societies and non-decimally in others. Despite intently looking for several decades now, the cognitive sciences have failed to find brain-bound mechanisms that can explain such differences. Assuming the brain does it all has kept researchers from looking "outside the box" at non-neurocentric mechanisms and resources.

In my work as a cognitive archeologist investigating numbers and writing, I use an extended-mind paradigm that sees material forms as substantially contributing to human cognition. The mind becomes a system that consists of brain, body, and material forms, components whose interaction is cognition. This paradigm alters the relation between the brain and material forms. While the brain remains a critically important part of the cognitive system, material forms no longer passively receive its externalized mental content. Instead, the material forms used to represent and manipulate numbers are integral to the conceptualization of numbers, and the incorporation of new material forms becomes the machinery of numerical elaboration. In this paradigm, cross-cultural variability in organization and extent becomes a simple matter of whether material devices are used, which ones are used, and how they are used (Overmann 2023). This "out-of-the-box" approach assumes the brain does not do everything unassisted.

Since a materially focused view of numbers is a radical upstart minority in researching numerical cognition, imagine my surprise—my shock of recognition, my excitement—when I first encountered the work of Cinzia Florio several years ago in an obscure proceedings published in Italian in 2009 (Florio 2009). *This is amazing*, I thought, *here is someone who thinks about numbers and material devices in the same way I do*. She understood that quipu knots were perfectly compatible with either positional or additive notation, the kind of insight I discover in my own work. She also talked about the three-dimensional device that was thought perhaps to have been used for calculating by the Inca, which—coincidentally—was the very device that first captured my interest in understanding the cognitive role of material devices in numbers. I contacted her, and I have been happily exchanging ideas with her ever since.

I read work about numbers across an interdisciplinary range of fields—psychology, linguistics, ethnography, philology, philosophy, mathematics, and history. In the literature, I find that when material devices are mentioned at all, their significant role in numerical conceptualization is not just simply ignored, it is not considered. But Ms. Florio's work puts the devices used by the Inca—the quipu for representing numbers, the yupana for manipulating them—at the very center of her investigations. She investigates the checkerboard yupana drawn in the early 1600s by an Indigenous author, Guamán Poma. She considers how this device worked hand-in-hand, both with the quipu and with verbal forms for numbers. She also decodes a plausible mathematical message on the yupana, a multiplication algorithm that suggests the Inca used calculating methods similar to those found in other mathematical traditions of the ancient world, particularly Mesopotamia, Egypt, and China. What unites these traditions is the logic of moving counters on a gridded surface.

Thinking "Outside the Box"

The novelty Ms. Florio brings to her interpretation of the Inca yupana starts with the recognition of the difference between Western numbers and the historical ones she investigates. From my own work in cultural number systems (Overmann 2025), I can affirm that learning to see and then set aside what our own numerical tradition has taught us about numbers is a difficult challenge. For example, we have been taught that when we count things by pairs, the counting goes: *two, four, six, eight, ten, … twenty*. In Polynesia, in comparison, historical counting by pairs proceeded: *one, two, three, … ten*, and the count was understood as doubled. Seeing the difference and setting aside the expectation enabled me to reconstruct the historical counting method used in Polynesia and trace its effect on the words for numbers recorded by early Western explorers (Overmann 2020). As another example, we have been taught that each next number is one more than the number preceding it, sometimes called the successor principle. We think that *eighteen* is *one* more than *seventeen* and *one* less than *nineteen*, and so we expect that this is the way numbers work. Numbers in other cultural number systems, however, can be based on a very different logic. In Kakoli, a language of Papua New Guinea, *eighteen* is *two* [in the next group of four above *sixteen*] toward *twenty* (Bowers and Lepi 1975). Setting aside the successor principle allows us to appreciate the fascinating beauty of numbers that work on different principles.

Similarly, in the case of the Inca numbers, most researchers have assumed the quipu and the yupana share the positional organization of Western numbers. Ms. Florio, recognizing this as a learned preconception about numbers, was able to set aside her Western education and training, and instead see the Inca devices on their own terms. This enabled her to recognize the quipu and yupana as representing additive numbers, rather than positional ones. This insight makes a lot of sense when we consider the way notational systems emerge and change: positional systems emerge from additive ones through lengthy processes of use and refinement involving material forms (Chrisomalis 2010). They do not spring forth from the unassisted brain, no matter how creative those brains might happen to be.

Ms. Florio was able to discard a second preconception having to do with *zero*, an essential component of the Western system of place value. While abacuses of any type, including the yupana, have an implicit *zero* in their design—the null condition in which a counter is not placed or a bead is not moved—*zero* need not be an explicit concept. In practical terms, this means that the notational system that accompanies an abacus need not incorporate an explicit *zero*. A paradigmatic example of this is the Roman abacus, a device whose calculations were recorded with numerals that famously lack a *zero*. Similarly, Chinese numbers were manipulated on abacuses of various designs or with counting rods, but Chinese numerals are multiplicative. In the multiplicative notations, a numeral does not assume the weight of, say, hundreds because it occupies the third position; rather, a numeral is specified as hundreds by a symbol that follows it to mark it as meaning hundreds. The salient point is this: the use of an abacus—including the yupana—neither determines nor demonstrates the presence of an explicit concept of *zero*.

Yet another novel aspect to Ms. Florio's yupana interpretation is her recognition of the fundamental difference between representing and manipulating. *Representation* displays numerical information, while *manipulation* moves it around. The devices performing these functions tend to be good at one or the other, but not both (Overmann 2016). For example, a tally displays numerical information through the quantity and organization of its notches, but because notches cannot usually be moved or removed once they have been made, the tally is not suitable for manipulating the numerical information it contains. Conversely, an abacus is good at moving numerical information around, but it cannot perform this function as long as it displays any particular number. Societies typically respond to this circumstance by using different devices to represent and manipulate their numbers. For example, the Egyptians, Mesopotamians, and Chinese used written notations for representing and pebble counters, clay tokens, or counting rods for manipulating their numerical information.

Within any particular number system, the representing and manipulating devices also work together. This principle can be illustrated with Roman numerals and the Roman abacus. Roman integers are decimal with a sub-base of 5 as reflected by symbols for 5 (V), 50 (L), and 500 (D). The abacus design incorporates the sub-base by dividing the columns of beads into groups of four and one. Calculation is a matter of moving beads to input the initial numbers and then manipulating the beads as needed for the calculation. The result of the calculation is then directly read from the beads that remain pushed toward the abacus centerline and directly written as Roman numerals (Pullan 1968).

The devices for representing and manipulating work together, not merely because both reflect the same number system, but also because their designs would have converged over time to make the transfer of information between them as easy and as accurate as possible (Florio and Overmann 2025). While this insight might seem to be something that should go without saying, it is generally overlooked with respect to Inca numbers, where the counterparts to the Roman devices are the quipu for representing and the yupana for manipulating numerical information.

Where previous researchers interpreting the yupana have often ignored the decimality of the quipu or the Inca spoken numbers, Ms. Florio instead begins her analysis by applying the two principles to the Inca devices. The quipu and yupana conform to the first principle—the fundamental difference between representing and manipulating—because they divide the functions between them. Simply, the quipu represents, and the yupana manipulates. The second principle holds that the representing and manipulating devices should work together. In this new work, Ms. Florio shows how and why the two work with each other, as well as with the decimal Inca number-words.

A final aspect of novelty is Ms. Florio's ability to consider how calculating is performed materially, by means of moving counters around on a gridded surface. The multiplication algorithm she decodes is plausible, not only because it is numerically consistent (which most interpretations are, no matter how farfetched in terms of the concepts they imply), and not only because it finds a numerical message (an important indication of decipherment validity). Rather, the algorithm she finds is recognizably similar to those of comparable numerical traditions, by which I mean numerical traditions that perform their calculations with some kind of abacus: moveable counters on a flat surface. The Inca multiplication method she finds is similar to—but more efficient than—an Egyptian method attested in the Rhind Mathematical Papyrus, a treatise dated to the mid-second millennium BCE (specifically, the method used to multiply 12 × 12 in Problem 32; see Damerow 2010; Reimer 2014).

If the Egyptian method were used to multiply 32×5, the multiplication Ms. Florio finds on the Inca yupana, the numbers would first be placed at the head of two columns, and then they would be doubled, selected, and added. While this method, like the Inca one, is based on the logic of moving physical counters, there are two major differences between the Egyptian method and the Inca method depicted on Guamán Poma's yupana. First, the elements encode their value differently. For Egyptian notations, encoding involves character shape; for the yupana, encoding involves kernel color. Second, the Egyptian algorithm mechanically doubles each row and then selects only those rows whose values add to the multiplicand; more rows are produced than are used. The Inca algorithm, in comparison, is more succinct, as it parses the multiplicand in a way that eliminates the formation of unused rows and reduces the need to place and move counters.

It is worth highlighting some of her other insights. One is that the yupana drawn by Guamán Poma was sized for a specific calculation, so it did not necessarily always have five rows and four columns. We tend to overlook this fluidity of form in making our own pencil-and-paper calculations, where, for example, dividing 23 by 4 requires fewer rows and columns than does dividing 9477 by 41. Another insight is that multiple yupanas might have been conjoined to perform calculations that were complex and lengthy. This same idea has been proposed for Mesopotamia by Christopher Woods, Professor of Assyriology and Sumerology, based on characteristic errors that occur midway in calculations of the Old Babylonian period, perhaps representing errors made in moving counters between two conjoined boards (Woods 2017). Such connections suggest there is a good potential for further research in these areas.

The importance of Ms. Florio's contribution should not be underestimated. Not only does she discard Western preconceptions about numbers to take the Inca yupana and quipu on their own terms, she also understands how they work as material devices, and accordingly, how they would have influenced Inca calculation. In the process of deciphering this historical artifact, she also demonstrates a powerful technique for understanding historical mathematics that can be applied to any cultural number system: the discarding of preconceptions, the consideration of the material devices and material logic used, and the synthesis of information from history, linguistics, mathematics, and culture.

Director, Center for Cognitive Archaeology
Anthropology Department
University of Colorado, Colorado Springs, USA
21 July 2024

Karenleigh A. Overmann

References

Bowers, N., & Lepi, P. (1975). Kaugel Valley systems of reckoning. *Journal of the Polynesian Society*, *84*(3), 309–324.
Chrisomalis, S. (2010). *Numerical notation: A comparative history*. Cambridge University Press.
Damerow, P. (2010). *Abstraction and representation: Essays on the cultural evolution of thinking*. Kluwer Academic.
Everett, D. L. (2005). Cultural constraints on grammar and cognition in Pirahã: Another look at the design features of human language. *Current Anthropology*, *46*(4), 621–646.
Florio, C. (2009). Encuentros y desencuentros nella individuazione di una relazione matematica nella yupana in Guamán Poma de Ayala. In E. Guagliano (Ed.), *Incontri e disincontri tra Europa e America, Atti del XXX Convegno Internazionale di Americanistica Salerno, 14–15 maggio e 10–12 dicembre 2008* (pp. 151–186). Oèdipus Editore.
Florio, C., & Overmann, K. A. (2025). Inka numbers, *khipu*, and *yupana*: A reanalysis. *Journal of Cognition and Culture*, *25*(1–2), 128–158.
Overmann, K. A. (2016). The role of materiality in numerical cognition. *Quaternary International*, *405*, 42–51.
Overmann, K. A. (2020). The curious idea that Māori once counted by elevens, and the insights it still holds for cross-cultural numerical research. *Journal of the Polynesian Society*, *129*(1), 59–84.
Overmann, K. A. (2023). *The materiality of numbers: Emergence and elaboration from prehistory to present*. Cambridge University Press.
Overmann, K. A. (2025). *Cultural number systems: A sourcebook*. Springer International Publishing.
Pullan, J. M. (1968). *The history of the abacus*. Hutchinson & Co.
Reimer, D. (2014). *Count like an Egyptian: A hands-on introduction to ancient mathematics*. Princeton University Press.
Woods, C. (2017). The abacus in Mesopotamia: Considerations from a comparative perspective. In L. Feliu, F. Karahashi, & G. Rubio (Eds.), *The first ninety years: A Sumerian celebration in honor of Miguel Civil* (pp. 416–478). Walter de Gruyter.

Preface

This book is the culmination of a study that I began about 20 years ago. This study started almost by accident, the result of a series of events. These put me on a path that, as I look back on it today, seems almost incredibly obligatory—perhaps even foreordained. This path led me to meet (or rather, collide with) a fascinating drawing: Guamán Poma's (1615) *yupana*. My great passions for archeology and mathematics did the rest.

The image in question, Guamán Poma's *yupana*, was drawn by an Inca nobleman around 1600, the early period of the Spanish colonization of Latin America. It depicts a calculation carried out using the mathematics of the ancient Incas. The drawing was part of a voluminous text, *El Primer Nueva Corónica y Buen Gobierno*, that somehow made its way into the Royal Library of Copenhagen, only to be forgotten and overlooked until it was rediscovered in 1908.

Since its rediscovery and availability for research, many scholars have tried to decipher the *yupana* as an Inca calculating device. To date, my work remains the only one that provides a decipherment, consistent with the very concept of deciphering. I identify a multiplication, and therefore, an algorithm that binds and connects the deciphered numbers with an evident mathematical logic. Other scholars attempting decipherment, in comparison, have identified only random numbers—numbers that are not connected to each other through any mathematical logic.

The subtitle I chose for the book—*A Hundred Years' Riddle*—reflects an incredible coincidence: the time that elapsed between the rediscovery of the *Nueva Corónica* in 1908 and the presentation of my deciphering at the XXX International Conference of American Studies in Salerno, Italy, in 2008 is exactly 100 years. The proceedings of the conference were published the following year, in 2009. As I write this preface, another scientific publication has just been published, an article in the *Journal of Cognition and Culture* that I co-authored with Prof. Karenleigh A. Overmann, a cognitive archeologist who investigates the materiality of numbers and writing systems (Florio and Overmann 2025). Our article reports my decipherment, and we frame the Inca numerical system and method of calculation in the context of the numerical systems and calculation methods of other ancient civilizations, finding analogies and coherence.

Beyond its explicit and obvious goal of reporting the decipherment, this book has a broader goal: It is intended to capture the interest not only of mathematicians and those who love mathematics and mathematical puzzles as a hobby but also the wider audience consisting of those who love and study Andean culture both academically and personally. The second group, which often includes people who enter into their studies with a strong focus and background in the humanities, has hitherto been deprived of the possibility of fully understanding the culture they love. This is because it is possible to study every little detail of a culture—its language, poetry, music, laws, history, traditions, customs, and religion—without fully grasping its technology and mathematics.

In academic research, humanistic and scientific studies are considered different pursuits. This separation creates a strong fracture that becomes evident when studying an ancient civilization like that of the Incas. But the culture of a people cannot be taken only halfway; it is incomplete if it does not include its mathematics and technology. In the same vein, mathematicians and scientists cannot approach the mathematics and technology of an ancient people without also entering into its traditions, its culture, its history, and its territory, because all these variables influence the evolutionary processes leading to technical resolutions of the problems faced by particular peoples in specific contexts.

For this reason, this book has been written in a simple and fluent way. The style is intended for both mathematical and non-mathematical readers. Mathematical readers are provided with all the information available from colonial texts and archeological finds, a body of knowledge that is more typically the domain of humanists. Non-mathematical readers are taken by the hand and provided with the conceptual tools necessary for understanding and solving the decipherment problem. The topic is developed gradually. Each step builds on the previous one and opens the way for the next, in order to lead readers to decipher the *yupana* as if they had done it themselves. This enables readers to understand—deeply—the logical paths once followed by the Incas.

The text also provides examples of calculations with detailed explanations. It also points out that there is another page in *Nueva Corónica* that seems to provide information about yupana, and this page has not previously been analyzed from this point of view.

My hope is that both kinds of readers—the mathematically inclined and the cultural enthusiasts—will enjoy exploring the mathematical world of the ancient Incas.

Moliterno (PZ), Italy Cinzia Florio
7 February 2025

References

Florio, C. (2009). Encuentros y desencuentros nella individuazione di una relazione matematica nella yupana in Guamán Poma de Ayala. In E. Guagliano (Ed.), *Incontri e disincontri tra Europa e America, Atti del XXX Convegno Internazionale di Americanistica Salerno, 14–15 maggio e 10–12 dicembre 2008* (pp. 151–186). Oèdipus Editore.

Florio, C., & Overmann, K. A. (2025). Inka numbers, *khipu*, and *yupana*: A reanalysis. *Journal of Cognition and Culture*, 25(1–2), 128–158.

Guamán Poma, F. de A. (1615). *El primer nueva corónica y buen gobierno*. Museum Tusculanum Press.

Acknowledgements

I want to infinitely thank Professor Rosa Maria Grillo of the University of Salerno. Back in 2008, despite the fact that her academic training was very far from mathematics, she dedicated herself to reading my work on the deciphering of Guamán Poma's yupana, and she noticed what she called an "overwhelming logical sequence." I was an unknown independent researcher, and despite this, she invited me—in my opinion, courageously—to report my research to the XXX International Conference of American Studies in Salerno. The publication of the proceedings of the conference in 2009 allowed me to start making my work known.

In the same way, I want to thank Don Mario Ferrari, Director of the "Ugo Morin" Educational Research Center. In 2011, he gave me the opportunity to publish an excerpt from the proceedings of the XXX International Conference of American Studies in the Italian journal *L'insegnamento della Matematica e delle Scienze Integrate* (*The Teaching of Mathematics and Integrated Sciences*). This begin to make my work known to mathematicians.

A deep thank you goes to Professor Karenleigh A. Overmann of the University of Colorado, Colorado Springs. Without her, this book would never have been written. Many years after the publication of the proceedings of the XXX International Conference on American Studies, while surfing the Internet, Professor Overmann came across my little-known and difficult-to-find work in the sea of academic publications. She contacted me, pushed me, and encouraged me to publish my discovery in an international specialized journal and to write a book. She supported me in translating my ideas into English and has done me the honor of writing a foreword for the book. All my immense gratitude goes to her.

Finally, I thank the people who have been close to me, who even if they do not love the history of mathematics and science, but love me. They have supported me by becoming involuntary and forced participants in this work. They have gladly given up the time and attention that I could, and perhaps should, have given to them, to do something that made me happy. I therefore thank my husband, Raffaele, and my children, Eliana and Gian Giacomo, for their support. In particular, I thank Eliana for the images she developed for the book.

Competing Interests The author has no competing interests to declare that are relevant to the content of this manuscript.

Contents

List of Figures

List of Tables

Chapter 1
Introduction

Abstract This chapter introduces the Inca, the devices used for calculating and recording numbers known as the quipu and yupana, the Chronicle written in 1615 by Guamán Poma that contains a drawing of a checkerboard yupana that is the key to the ancient Inca method of calculation.

Keywords Quipu · Yupana · Chronicles · Guamán Poma · Decipherment

All past civilizations with a high level of social organization had necessarily developed a spoken language, some method of recording language, and a more-or-less elaborate mathematical language. Today, we speak many different languages and dialects, but most societies use the same written numbers: the decimal positional numeral system with the digits 0–9 known as Western numbers. We also use the same algorithms—computational procedures that allow us to add, subtract, multiply, and divide. Beyond addition, even more complex algorithms allow us to determine the value of roots, logarithms, derivatives, integrals, etc.

In other words, while a word like "sun" is pronounced and written in many different ways across the world, the multiplication 123 × 456 is written in much the same way—perhaps even with the same notations—and carried out with very similar algorithms.

The unification of mathematical language was perhaps one of the first forms of globalization. While this unification has enabled us to understand each other better mathematically, we have also now forgotten the many different mathematical notations and algorithms once developed by ancient civilizations. Some have been completely lost at this point.

At first glance, the loss might seem inconsequential. Certainly, we are able to perform complex calculations with our own powerful numerical system and algorithms—so, why should we care how the Egyptians, Hittites, or Incas counted and calculated?

To understand the deep and hidden importance of an ancient numerical system and its algorithms for calculating, we must go beyond the simple need to obtain a numerical result. Behind the orderly structures that may seem boring to some, there

C. Florio, *The Mathematical Enigma of Guamán Poma's Yupana*,
https://doi.org/10.1007/978-3-031-92662-4_1

is intelligence, creativity, logical rigor—even genius. This is the spark that distinguishes our species from all other animals on the planet.

Getting our hands on the numerical systems and calculation methods of ancient peoples gives us a feeling of immersing ourselves in the ancient human mind and its evolution. We perceive the origins of our intelligence—our human roots—when we follow the reasoning, the logical paths created to find the right path, the winning strategies once used in ancient times.

It is exciting to observe how, many centuries ago, the Egyptians, Romans, and Chinese carried out the same multiplication—not just using different tools and symbols, but also different logical paths—to arrive at the same result. These algorithms have some similarities and points of contact, but they are all absolutely original and unique. They are so different from each other, despite being aimed at the same goal, and so brilliantly conceived, as to yield an immediate and spectacular picture of the immense potential of the human mind. Observing them and following their logic can generate emotions not too different from the enchanted awe we might feel when we look at a painting and feel a sense of happiness without understanding why; or we realize we are crying while listening to music that seems to be in tune with every particle of our body.

An example of using radically different tools, symbols, and logical paths to achieve the same result is shown in Fig. 1.1: the multiplication 28 × 15 = 420 performed with Egyptian and Chinese algorithms. Next to it are the corresponding numbers used today. For now, there is no need to worry about understanding the methods or their differences; it is enough to recognize that the methods are completely different, and yet they yield the same result.

Every society should be proud of its mathematical roots—in the same way it is proud of its poetry, its music, its art—because the mathematics of a people is an expression of its intelligence, its capacity for analysis, elaboration, and abstraction. Every society has the right to know its mathematical roots, and the peoples of the modern Andean states are among those who have the greatest right to it. This is

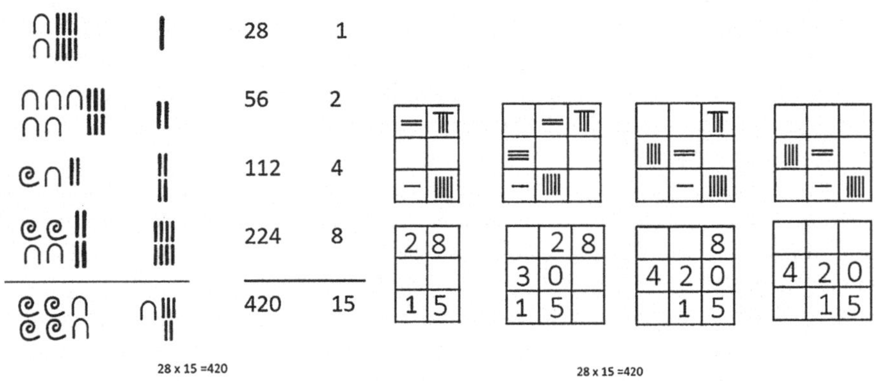

Fig. 1.1 Methods of multiplying 28 × 15. (Left) Egyptian notational algorithm; (right) Chinese counting rods algorithm. Both yield 420 as the product. (Image by the author)

because the descendants of the Incas did not lose their ancient mathematical knowledge due to a slow evolution and transformation, but rather, due to traumatic *force majeure*: the Spanish Conquest that eliminated or modified much of the Inca culture.

We would know little about this ancient people today if the same Conquerors—the secular and religious authorities who converted the Incas to Christianity, along with some Hispanicized Inca descendants—had not produced texts on their customs and traditions. This body of literature is known as the "Spanish Chronicles." These texts tell us about the life of the Incas in detail, but when it comes to topics such as how the Incas calculated or recorded their numbers or words, no Chronicle is exhaustive. This was probably not because the various Chroniclers intended to obscure the methods they observed, but more likely because they did not understand any means of recording other than the one they were familiar with: writing symbols on paper with pen and ink. In fact, the Incas used an ingenious alternative means to record their numerical and narrative information: groups of colored threads tied with knots like the one shown in Fig. 1.2. These devices are known as "quipus" after the Quechuan word for "knot."

The Conquistadors were Europeans who lived between the sixteenth and seventeenth centuries. They came to the New World ready and quite determined to label everything done by the natives as "primitive" and "useless." This was certainly their reaction to Inca writing and mathematics. As far as the latter is concerned, there is another factor to consider: in Europe at the time, "calculating" was not a widespread skill. Rather, it was an activity performed almost exclusively by specialized accountants, and they tended to keep their techniques to themselves. In addition, the historic methods of calculating with abacuses (or counting boards with counters) and recording the results with Roman numerals were in the process of being replaced by a new way of calculating with algorithms involving a new kind of written notations. Unlike Roman numerals, these new notations were positional numbers, weighted by

Fig. 1.2 Quipu (Locke 1912, pl. XXII). (Image in the public domain)

place value. While they were new to Europe at the time, they are familiar today as the Western numerals so widely used around the globe. The new "algorist" method of calculating with the new notations was even less well-known to the common people than the previous "abacist" method had been, and it introduced another revolutionary concept—"zero."

The Conquistadors and their Chroniclers were men who most likely called on an accountant if they needed any calculation done. As such, they were hardly in a position to have explained how the Incas calculated—let alone to have compared it to the Western method in a way we would find useful today. They simply lacked the knowledge and motivation needed to do so. Unfortunately, no expert accountant of the time who could have explained Inca mathematics in detail appears to have been interested enough in the Inca method of calculation to record his observations of the matter. Such a record is simply not known to exist.

To make an unfortunate situation even worse, the clergy destroyed many quipus and discouraged their use. As a result, shedding light on traditional Inca mathematics has been a slow and ongoing process. This text aims to retrace the phases of reconstruction. It also aims to insert the last pieces needed to complete the puzzle of Inca mathematics and thus return the mathematics of their ancestors to the Andean people.

This work is innovative because it identifies—both on the quipu and the yupana (the Inca calculating table)—an additive decimal system like Roman and Egyptian numerals, rather than a positional decimal system like Western numerals. The work also reads a mathematical message on the yupana drawn by Guamán Poma, the only example of the use of the yupana known today. The message read there is a multiplication algorithm, albeit one with a mistake in the drawing, a single wrong symbol out of a total of 55. All other studies of Guamán Poma's yupana have identified only random numbers, ones not connected to each other by a mathematical logic or message, and which therefore lack convincing internal proof. (As will be explained later in detail, "without proof" is equivalent to saying that there could be 55 errors out of the total of 55 symbols.) Finally, this work gives a more complete view of the form of the yupana.

This book has been written as a succession of small logical steps. They are intended to be clear and simple so that readers are guided and supported in making the connections between them. The ideas will hopefully be accessible to everyone—even those who may not love mathematics.Competing InterestsThe author has no conflicts of interest to declare that are relevant to the content of this chapter.

Reference

Locke, L. L. (1912). The ancient quipu, a Peruvian knot record. *American Anthropologist, 14*(2), 325–332.

Chapter 2
Quipus Recounted in the Chronicles and the Studies to Decipher them

Abstract This chapter illustrates the quipu as seen in colonial sources, shows the history of the decipherment of the numerical quipu, and introduces a doubt about the number system as deciphered by Locke.

Keywords Quipu · Numerical quipu · Decipherment · Chronicles

The Chronicles often mention quipus, and they describe them as consisting of a main string from which many smaller colored strings (pendants) hung like a fringe; other, subsidiary strings could also be tied to the pendants. Knots tied into the strings could represent numbers (numerical quipus), or they could indicate information or concepts (narrative or historical quipus). Concerning the latter, however, the Chronicles are often contradictory or at least unclear: sometimes they speak of narrative quipus as objects comparable to books, as if the Inca were writing in knots; at other times, they describe narrative quipus as mnemonic objects, like rosaries used for prayers, supporting the recollection of information known by heart, like a poem.

Here are some quotes from the Chronicles:

Pedro Sarmiento de Gamboa, *Segunda Parte de la Historia General Llamada Indica*, Cuzco, 1572.

> … para suplir la falta de letras, tenían estos bárbaros una curiosidad muy buena y cierta... Y así cada uno á sus decendientes iba comunicando sus anales por esta orden dicha, para conservar sus historias y hazañas y antigüedades y los números de las gentes, pueblos y provincias, días, meses y años, batallas, muertes, destruiciones, fortalezas y cinches. Y finalmente las cosas mas notables, que consisten en número y cuerpo, notábanlas, y agora las notan, en unos cordeles, á que llaman quipo, que es lo mesmo que decir racional ó contador. En el cual quipo dan ciertos ñudos, como ellos saben, por los cuales y por las diferencias de las colores distinguen y anotan cada cosa como con letras. Es cosa de admiración ver las menudencias que conservan en aquestos cordelejos, de los cuales hay maestros como entre nosotros del escrebir. (De Gamboa 1572, p. 31)

> … to make up for the lack of letters, these barbarians had a very good and certain curiosity... And so each [man] communicated his annals to his descendants by this said order, to preserve his histories and deeds and antiquities and the numbers of the people, towns and provinces, days, months and years, battles, deaths, destructions, fortresses and "Sinchis" [regional warlords]. And moreover, the most important things, which concerned numbers

C. Florio, *The Mathematical Enigma of Guamán Poma's Yupana*,
https://doi.org/10.1007/978-3-031-92662-4_2

and materials, they wrote down, and still write down on strings that they call quipu, which is the same as saying rational or counter. In these quipus they give certain knots, as they know, by which and by the differences of the colors they distinguish and write down each thing as with letters. It is a matter of wonder to see the details they preserve in these cords, of which there are masters of writing, as among us. (as translated by the author)

José de Acosta, *Historia Natural y Moral de las Indias*, Seville, 1590.

Son Quipos unos memoriales, o registros hechos de ramales, en q' diversos ńudos, y diversas colores significá diversas cosas. Es increible lo q' en este modo alcançaró, porq' quáto los libros puedé dezir de historias, y leyes, cerimonias y cu é tas de negocios, todo esso suplé los Quipos tan puntialmente, q' admira. (De Acosta 1590, Book VI, Chapter VIII)

Quipus are memory aids or registers made up of cords on which different knots and different colors signify different things. What they achieved in this way is incredible, for whatever books can tell of histories and laws and ceremonies and accounts of business all is supplied by the quipus so accurately that the result is astonishing. (as translated by Martin and Wasserman 2010, p. 5)

Garcilaso de la Vega, *Commentarios Reales, que Tratan del Origen de los Yncas*, Lisbon, 1609.

A cierto tiempo seńalado se juntauá er el pueblo principal de ca da prouíncia los juezes cobradores, y los cótadores, o escribanos q' teniá los ńudos y cuentas de los tributos: y delante del Curaca y del Gobernador Inca hazíá las cuétas, y partíciones por los ńudos de sus hilos, y con piedrezuelas cóforme al numero de los vezinos de latal prouíncia, y las sacauá tan ajustadas y verdaderas, q' en esta parte yo no se a quien se pueda atribuyr mayor alabança, si a los contadores q' sin cífras de guarismos haziá sus cuentas y partíciones tá ajustadas de cosas tan menudas, cosaq' nuestros Arítmetícos suelé hazer có mucha díficultad: o al Gouernador y mínístros regíos, q' con tanta facílídad entédian la cuenta y razon, que de todas ellas les dauan. (De la Vega 1609, Book II, Chapter XXVI)

At an appointed time the judges responsible for the collection and accountants or scribes who kept the knots and beads for reckoning the tribute assembled in the chief town of the province: the calculations and divisions were made in the presence of the curaca and the Inca governor by means of the knots on the cords and small stones, according to the number of householders in the province. The calculations were so exact that I hardly know whether to praise the more the accountants who made their reckonings without the use of figures and contrived to divide exactly very small quantities, a thing our mathematicians have great difficulty in doing, or the royal governor and officials who followed the process with perfect ease. (as translated by De la Vega 1966, pp. 274–275)

Garcilaso de La Vega, *Commentarios Reales, que Tratan del Origen de los Yncas*, Lisbon, 1609.

Pero lo que contenia la embaxada, ni las palabras del razonamiento , ni otro suceslo historial, no podia dezirlo por los nudos: por que consiste en oracíon ordenada de víua voz, o por cscrito, la qual no se puede referir por nudos, por qel nudo díze el numero, mas no la palabra. (De la Vega 1609, Book VI, Chapter IX)

But the purpose of the embassies or the contents of the speeches, or any other descriptive matter could not be recorded on the knots, consisting as it did of continuous spoken or written prose, which cannot be expressed by means of knots, since these can give only numbers and not words. (as translated by Livermore, 1966, pp. 331–332)

> Los ńudos se dauan por su orden de unídad, dezena, centena, millar, dezena de míllar, y pocas vezes o nunca, passauan a la centena de millar: por que como cada pueblo tentasu cuenta de por si, y cada metro poli la de su distríto, nunca llegaua el numero destos, o de aquellos a tan ta cantídad que passasse al cétena de millar, que en los numeros q' ay de ailí abaxo teniá harto. Mas si se osre ciera auer de centar por el numero centena de millar tambi/ en lo conta ran: por que en su lengua je pueden dar todos los numeros del guarismo como el los tiene mas por q' no auia para que usar de los numeros mayores no passaua del dezena de millar. Estos numeros centauan por ñudos dados en aquellos hilos, cada numero diuidido del otro: empero los ñudos de cada numero estauan dados todos juntos debaxc de una buelta, a manera de los ñudos que se dan en el cordon del bienauenturado Patriarca San Francisso, y podí asse hazer bien por q' nunca pasiauan de nueve como no passan de nueve las unidades y dezenas, &c. En lo mas alto de los hílos poníá el numero mayor, que era la dezena de millar, y mas abaxo el millar, y assi hasta la unídad. Los ńudos de cada numero, y de cada hílo yuan parejos unos con otros, ní mas ni menos q' los pone un buen contador, para hazer una suma grande. (De la Vega 1609, Book VI, Chapter VIII)

> The knots were arranged in order of units, ten, hundreds, thousands, tens of thousands, and seldom if ever passed a hundred thousand, since as each village kept its own records, and each capital the records of its districts, the numbers never in either case went beyond a hundred thousand, though below that figure they made many calculations. If they had had to count hundreds of thousands, they would have done so, since their language has words for all possible numbers known in arithmetic. But as they had no cause to use larger numbers, they did not go beyond tens of thousands. These numbers were reckoned by means of knots in the threads, each number being divided from the next. But the knots representing each number were made in a group together, on a loop, like the knots found in the cord of our blessed patriarch St. Francis: this was not difficult to do, as there were never more than nine, seeing that units, tens, etc., never exceed nine. The greatest number, say tens of thousands, was knotted at the upper end of the threads, the thousands lower down, and so on down to units. The knots for each number on each thread were exactly alike, precisely as a good accountant sets his figures to make a long addition. (as translated by Livermore, 1966, pp. 331–332)

As for the quipus on which accounting data were recorded, the numerical quipus, the most detailed testimony is perhaps that of Garcilaso de la Vega (precisely the last on the list) who had the opportunity to use them in his youth, as he recounts in his Chronicle. But even this explanation is not completely clear, so much so that it was misunderstood when it was read in the first half of the XVIII century by Raimondo di Sangro, Prince of San Severo, the first scholar to take an interest in numerical quipus. Although the prince was a very cultured man of marked intelligence, he did not have a numerical quipu to observe, only the words written by Garcilaso. As a result, he drew the quipu shown in Fig. 2.1.

In his diagram, San Severo had wanted to represent the number 21,314, but he did not succeed. This was clear after the successful decipherment of the numerical quipus made at the beginning of the twentieth century by Leslie Leland Locke. San Severo's quipu contained 11 knots that are read, from top to bottom, as two Franciscan knots with five turns, one Franciscan knot with four turns, three Franciscan knots with three turns, one Franciscan knot with two turns, and four simple knots. As shown in Fig. 2.2, the knots, depending on the power of 10 they represented, had a different shape:

Fig. 2.1 Quipu as drawn by Raimondo di Sangro, Prince of San Severo (Di Sangro Sansevero 1750, p. 193). (Image in the public domain)

The desired number was obtained by adding the values associated with each knot on the string. It is apparent that knotting this string would have required considerable time and effort.

About 400 years after the Conquest, Leslie L. Locke deciphered the numerical quipus in 1912 by referencing Garcilaso's Chronicle and using quipu B8713 (Fig. 2.3) from the collection of the American Museum of Natural History.

The same quipu, as diagrammed by Locke, is shown in Fig. 2.4:

B8713 consists of six distinct groups of cords. Each group is tied to the main cord and consists of four pendant cords. In addition, each group has a cord arranged

Knot shape	Value
 simple knot	1
 Franciscan knot with two turns	10
 Franciscan knot with three turns	100
 Franciscan knot with four turns	1000
 Franciscan knot with five turns	10000

Fig. 2.2 The values assigned to the knots in the quipu as hypothesized by the Prince of San Severo. (Created by the author using images in the public domain)

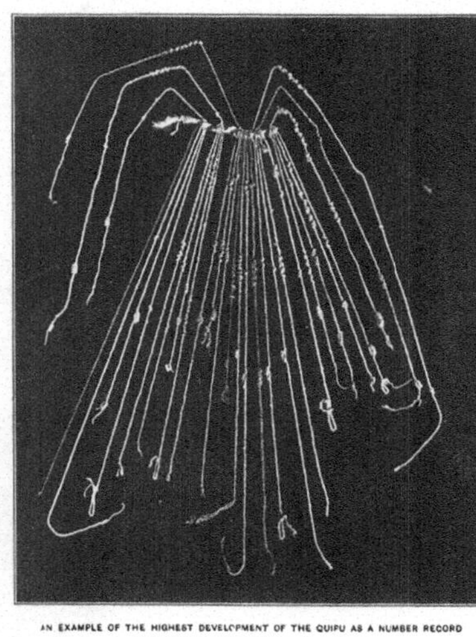

Fig. 2.3 Quipu B8713 (Locke 1912, pl. XXIV). (Image in the public domain)

in the opposite direction (an upper cord) that passes through the four loops hooking the pendant cords to the main cord (Fig. 2.5).

The presence of four strings for each group seems to indicate the need to account for the same four parameters, but for six different components or in six different tempos.

> The more the *quipucamayu* considers the pattern of distribution by taking account of the relative sizes and positions of different cords, the better the logical structure of the final representation. Cord placement, color coding, and number representation are the basic constructional features, repeated and recombined to define a format and convey a logical structure. This search for a coherent numerical/logical structure is mathematical thinking. (Joseph 2011, p. 55)

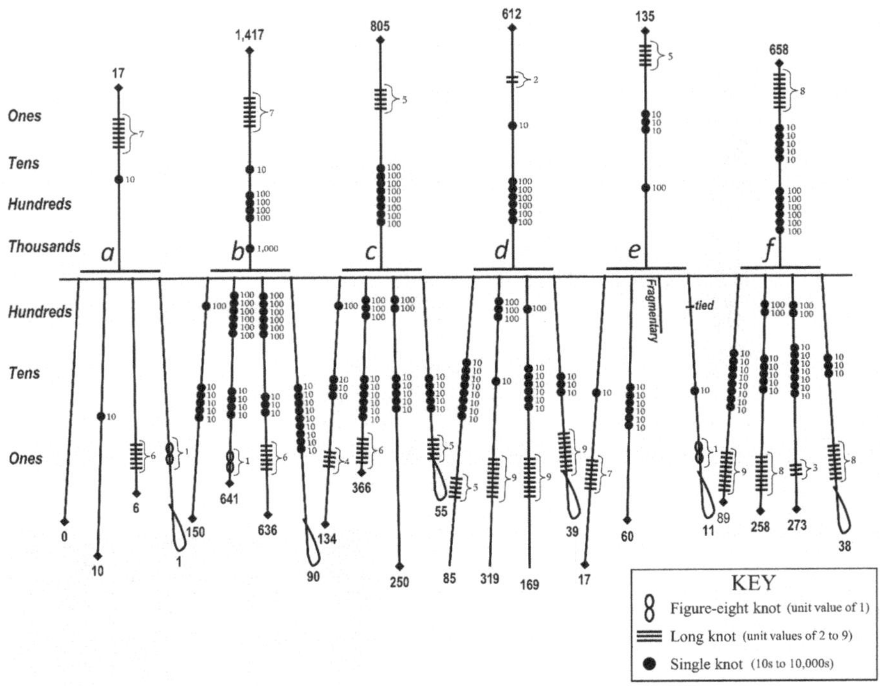

Fig. 2.4 Diagram of quipu B8713. (Redrawn from Locke (1912, pl. XXIII) by KAO and previously published in Florio and Overmann (2025, p. 134, Fig. 4) under a Creative Commons license)

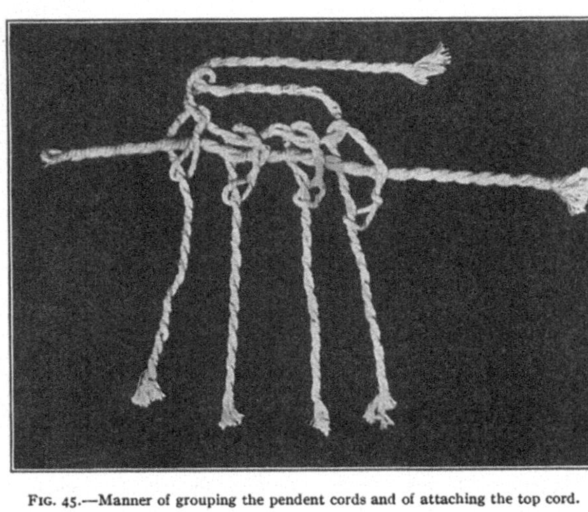

Fig. 2.5 B8713 construction (Locke 1912, Fig. 45, p. 329). (Image in the public domain)

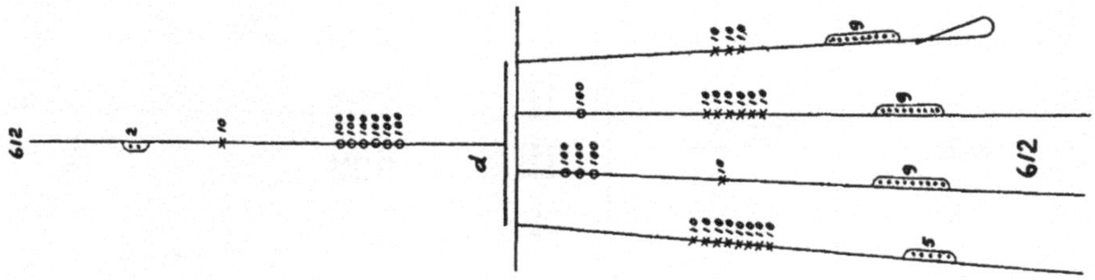

Fig. 2.6 Detail of quipu B8713 (extracted from Locke 1912, pl. XXIII). (Image in the public domain)

In Quipu B8713, for example, it could be assumed that the census of six communities was represented, detailing the four components: men, women, children, and the elderly.

Locke hypothesized that the knots, depending on the area in which they were located along the string, represented the values 1, 10, 100, etc. In this, he followed Garcilaso's association of knots with these values. He also noted that the knots were grouped along the strings in only some areas, and they were well spaced from each other. Moreover, these areas were repeated in parallel across the strings, creating bands. All of the knots at the ends of the strings were either figure-eight knots (Solomon's knot) or Franciscan knots with the number of twists ranging from two to nine. Locke reasoned the twists would indicate the numbers 1–9. The knots in the upper bands, which were all simple knots, were tied one next to another, and they varied in number from 1 to 9. Locke thought the area in which the knots were tied, from bottom to top, would indicate the tens, hundreds, etc. Locke wrote the assigned value next to each knot (Fig. 2.6).

Locke's interpretation could have remained hypothetical, and others could have been proposed as well (as for example, the knots might be valued with the lowest at the top of the string, assuming that Garcilaso had not indicated the opposite precisely). Other, more fanciful hypotheses could also have been made as well. This raises the question of why Locke's interpretation would be valid.

The answer is simple: Locke's interpretation is valid because it can be demonstrated with internal consistency. For B8713, when the values represented on the four strings of a group are summed, they equal the number displayed on the corresponding top cord. If Locke's hypothesis had been wrong, such internal consistency would not always be possible.

Consider, for example, the third group of four strings shown in Fig. 2.7.

According to Locke's proposal, the numbers represented are 134, 366, 250, and 55. Their sum is 805. Reading the upper cord in the same way, we find the number 805. If, on the other hand, we had reversed the order, we would have read the pendants as 431, 663, 52, and 550, numbers whose sum is 1696. The upper cord would now be read as 508, which clearly does not match. The same is true for the other groupings, though the fifth group is inconsistent because one of the four strings is broken. Further, there is an error in the second group because the sum of its strings yields 1517, and the upper cord gives 1417; a knot is missing in the area of the

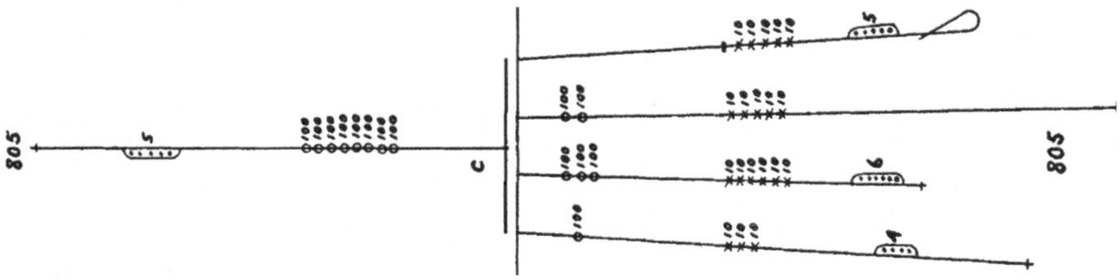

Fig. 2.7 Detail of quipu B8713 (extracted from Locke 1912, pl. XXIII). (Image in the public domain)

hundreds.[1] Only for the first group would reading in either direction produce consistent data, but only because the numbers involved are small and one of the strings is knotless. It is a special case, simpler, with fewer constraints.

In practice, Locke can support his hypothesis because he identifies an addition: a logical relationship that binds all the numbers involved, recognizing them as addends and sums. If there had been no upper strings on B8713, Locke could have offered the hypothesis, but not validated it. Without such validation, any other hypotheses, including the one inverting the values 1, 10, 100, etc. from top to bottom, would have the same validity as Locke's, no matter how fanciful.

In science, hypotheses are the soul of research. They push researchers to think, to find solutions, to persist when if the wrong path is initially taken. However, only hypothesis that can be validated is correct, and when this happens, all of the others are put away. The term "put away," in this context, does not mean "discarded," for in the history of science, wrong answers are useful and instructive, if not essential, because they show how a particular solution was found. They also reveal logical and creative minds looking for answers.

To this we can add that in the quipus, the knots representing the units (the Solomon's knot for 1 and the Franciscan knot with twists for 2–9) had been intelligently inserted by the Incas, differentiating the units by using knots of a different shape from the simple knots used in the higher areas. This practice made the units a fixed reference point on the strings that enabled the higher values to be identified and assigned the correct values.

This practice is reminiscent of the staff used to write music: the balls on the staff indicate different notes, depending on where they are located; similarly in quipus, knots indicate units, tens, hundreds, etc., depending on the area of the string on which they are performed. Thus, in music, the staff is an essential basis for writing notes; just so in quipus, the areas on the strings had to be well defined so the knots were correctly understood, and this could only be done with a reference point. In

[1] Perhaps there was no error, as one of the pendants contains the trace of a subsidiary string, presumably broken. The broken subsidiary might have contained the instruction to subtract a hundred. In this regard, Sabine Hyland's latest work, *Knot Anomalies on Inka Khipus: Revising Locke's Knot Typology* is illuminating (Hyland 2024).

fact, while in the staff there are always five lines, in quipus, the number of levels was not fixed; it could vary depending on the size of the numbers to be recorded on that particular occasion. If numbers on the order of a hundred were needed, only three levels were required; if numbers on the order of a thousand, four levels, and so on. In the absence of a fixed number of levels, the only solution was to set a reference level with which to compare the others. This was done with the units by diversifying the shape of the knots.

Locke's interpretation is also consistent with the maximum number of knots in each level, which is nine. In fact, no more than nine knots could be found per sector, as adding one would have increased to 10, but the value of 10 knots in one level corresponds to the value of only one knot in the next level (this information was also provided by Garcilaso).

In addition to the numerical type of quipu, quipus have been found with different characteristics. For example, some use Franciscan knots on all levels, or they have more than nine turns of string, or the color of the thread changes inside the knot. These quipus have not yet been deciphered, and they are thought to contain non-numerical information (narrative quipus).

Speaking of the techniques used by the Incas to transmit information, it is necessary to add a small remark about *tocapu*. These are mostly geometric symbols represented within small squares and reproduced both on clothes and on ceramics and petroglyphs (Fig. 2.8). Their origin predates the Inca culture, in fact they are found in many pre-Inca cultures.

Many studies have converged on the idea of value in the transmission of messages (a kind of writing), even if no certain results have been obtained. Among the best-known scholars in this line are Gail P. Silvermann (2014), Victoria de la Jara (1975), and William Burns Glynn (1981). I too became interested in the study of *tocapu*, but I followed an alternative path of investigation to identify a strong correspondence between *tocapu* and Chladni figures (the symbols of the *tocapu* would seem to be connected to the sounds) that needs further investigation to evaluate its validity, (Florio 2019).

Returning to the quipus, there is a final consideration, but one that is perhaps the most important: Locke, in interpreting the numerical quipus, uses a decimal system without specifying whether it is additive, like Roman numerals, or positional, like Western ones. Nonetheless, after his decipherment, everyone assumed the quipus represented a positional system, likely because the positional system was more familiar. Since positional systems require a "null" element (in Western numerals, this is the number zero), this element was identified not by its presence, but by its absence. That is, if there were no knots in a particular area of the string, then this indicated the presence of zero.

Needless to say, if—instead of the knotless stretch of string—there had been a knot of a particular shape to indicate zero, its presence would have been more convincing.

Fig. 2.8 Inca tunic with *tocapu*. (Photography by Neil Greentree © Dumbarton Oaks, Pre-Columbian Collection, Washington, D.C.)

The possibility of an additive system that does not require zero has never before been considered. On the other hand, the absence of a specific type of knot on the quipu to indicate zero (that is, the lack of any unique knot for representing zero) should have led to the strong suspicion that the numerical system was not positional, but additive. This will be one of the points that will be considered and demonstrated in the present work.

For readers who may be interested in the topic but not understand concepts like positional and additive numbers, it is necessary to give explanations. We will do this in the clearest and simplest way possible in the next chapter.

Competing Interests The author has no conflicts of interest to declare that are relevant to the content of this chapter.

References

Burns Glynn, W. (1981). La escritura de los Incas. Una introducción a la clave de la escritura secreta de los Incas [The writing of the Incas. An introduction to the key to the secret writing of the Incas]. *Boletín de Lima, N12–14*, 1–32.

De Acosta, J. (1590). *Historia natural moral de las Indias, en que se tratan las cosas notables del cielo, y elementos, mentales, plantas, y animales dellas: y los ritos, y ceremonias, leyes, y gouierno, y guerras de los Indios*. Juan de Leon.

De Gamboa, P. S. (1572). Segunda parte de la Historia general llamada Índica, la cual por mandado del excelentísimo señor Don Francisco de Toledo, virrey, gobernador y capitán general de los reinos del Pirú y mayordomo de la casa real de Castilla, compuso el capitán Pedro Sarmien. In R. Pietschmann (Ed.), *Geschichte des Inkareiches von Pedro Sarmiento de Gamboa (1906). Abhandlungen des Königlichen Gesellschaft der Wissenschaften zu Göttingen. Philogisch-Historische Klasse. Neue Folge, Band VL* (pp. 1–138). Weidmannsche Buchandlung.

De la Jara , V. (1975). *Introducción al estudio de la escritura de los Incas*. Studium.

De la Vega, G. (1609). *Commentarios reales, que tratan del origen de los Yncas, reyes que fueron del Peru, De su idolatria, leyes, y gouierno en paz y en guerra: de sus vidas y conquistas, y detodo lo que sue aquel Imperio y su Republica, antes que los Españoles passaran a el*. Pedro Crasbeeck.

De la Vega, G. (1966). *Royal commentaries of the Incas, and general history of Peru. Part one* (H. V Livermore (trans.)). University of Texas Press.

Di Sangro Sansevero, R. (1750). *Lettera apologetica dell'Esercitato Accademico della Crusca contente la difesa del libro intitolato Lettere d'una Peruana per rispetto alla supposizione de' quipu scritta alla Duchessa di S**** e dalla medesima fatta pubblicare*. Licenza de Superiori.

Florio, C. (2019). Il linguaggio dei tocapukuna per riscrivere la storia [The language of the tocapukuna to rewrite history]. In G. Nuzzo (Ed.),*"Letteratura testimoniale e costruzione della storia," atti del 40° Convegno Internazionale di Americanistica ["Testimonial literature and the construction of history," proceedings of the 40th International Conference on American Studies]* (pp. 49–67). Oédipus Editore.

Hyland, S. P. (2024). Knot anomalies on Inka khipus: Revising Locke's knot typology. In C. Orsini & F. Villa (Eds.), *IX Jornadas Internacionales de Textiles Precolombinos y Amerindianos, Museo delle Culture, Milan, 2022* (pp. 162–180). Zea Books.

Joseph, G. G. (2011). *The crest of the peacock: Non-European roots of mathematics* (3rd ed.). Princeton University Press.

Locke, L. L. (1912). The ancient quipu, a Peruvian knot record. *American Anthropologist, 14*(2), 325–332.

Martin, C. E., & Wasserman, M. (2010). *Writing and record-keeping in the ancient Americas: The quipus of the Andes. In Latin America and its people: Volume 1, To 1830* (2nd ed., pp. 1–13). Prentice-Hall.

Silvermann, G. P. (2014). *Los signos del imperio*. Biblioteca Abraham Valdelomar.

Chapter 3
Additive and Positional Numerical Systems

Abstract In this chapter, the concepts of additive and positional number systems and their differences are explained in a simple way, understandable even to experts in other fields (humanists in a general sense).

Keywords Quipu · Additive numbers · Positional numbers · Place value · Number systems

In an additive number system, the symbols used to express quantities—whatever their shape, whatever they are made of—always retain their value, regardless of their position in the context of the number, and the number is calculated by adding the values of all the symbols, one after another. For example, the ancient Egyptians used an additive number system with the powers of 10 (Fig. 3.1):

This number system is "additive" because addition is the only operation used to reach the total number, and each symbol can only be repeated up to nine times in representing any particular number. If a symbol has to be added for the tenth time, instead of using 10 identical symbols, they are all replaced by a single symbol for the next higher power of 10, which differs in representing the next higher quantity by 10. (Other additive number systems might use, for example, groupings of three or five symbols, and these are called additive number systems for the powers of 3 or 5, respectively.)

Consider, as an example, the number 1,231, which was written as follows in the Egyptian additive number system for the powers of 10 (Fig. 3.2):

If the symbols were rearranged like they are in Fig. 3.3, or if they were arranged in any other order, the number would remain unchanged. Nonetheless, the order could be criticized, as it was customary to group any repeated symbols together, and to place all of the symbols in an ascending or descending sequence; these customs made it visually easier to add the symbols to reach the total number and read the result immediately.

In an additive system, zero is not present, having no reason to exist: it makes no sense to add nothing, as the sum would always remain the same.

C. Florio, *The Mathematical Enigma of Guamán Poma's Yupana*,
https://doi.org/10.1007/978-3-031-92662-4_3

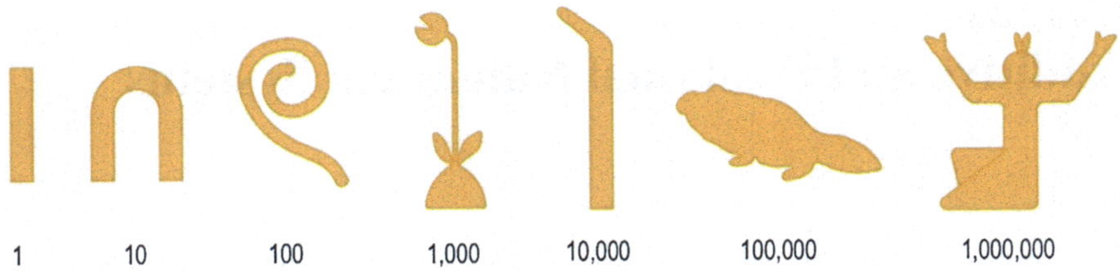

Fig. 3.1 The Egyptian numerals. Image by the author

Fig. 3.2 The number 1,231 written in Egyptian numerals. Image by the author

Fig. 3.3 The number 1,231 as rearranged. Image by the author

As a positional system, in comparison, we will consider the Hindu–Arabic system (or Western numerals) now used around the world. In this positional system, the symbols used to express quantities and which we call digits are 0, 1, 2, 3, 4, 5, 6, 7, 8, and 9. These symbols do not always represent the same value, as in the Egyptian system, but rather, their value changes depending on the position they occupy within the number (hence, the name "positional").

In fact, the quantities expressed by the digits, before they are added to produce the total number, are first multiplied by a term that differs according to the position in which the digits are located. This term is not always the same.

Let's consider again the number 1,231, which we previously used with the Egyptian system. In our base-10 positional system, this number is constructed as shown, starting from the right-most digit in the number, the first, rightmost digit (1) is in the units position; the second digit (3), the tens, the third digit (2), the hundreds; and the last, leftmost digit (1), the thousands. More formally, the rightmost digit is multiplied by 10^0 (i.e., 10 raised to the zero, which is equal to 1), and this yields 1; the second digit is then multiplied by 10^1 (i.e., 10 raised to one, which is equal to 10), yielding 30; the third digit is then multiplied by 10^2 (i.e., 10 raised to two, which is equal to 100), yielding 200; the fourth digit is multiplied by 10^3 (i.e., 10 raised to three, which is equal to 1,000), yielding 1,000. If the number were higher (say, 21,231), the sequence would continue in the same manner with the higher powers of 10.

All the numbers obtained are then added together to obtain the value of the number being represented: 1,000 plus 200 plus 30 plus 1 equals 1,231. Below are all the steps in numerical symbolism:

$$1\times10^3+2\times10^2+3\times10^1+1\times10^0=1\times1{,}000+2\times100+3\times10+1\times1$$
$$=1{,}000+200+30+1=1{,}231$$

The value 10 is called the base, and the number system is called the positional number system in base 10. (There are positional number systems with different bases as well, as for example, the well-known binary system is a base-2 positional number system that uses the digits 0 and 1.)

As can immediately be seen, the positional construction of the same number is more complex than it was in the additive Egyptian system. This is something we may not realize because we have been taught these numbers since childhood, and we use its conventions instinctively. Yet reading this number means performing nine operations instantly and simultaneously: four elevations to the powers of 10, four multiplications, and one final addition. In the additive system, in comparison, only one operation is needed, an addition.

Another difference is the use of zero, which a positional system requires to keep the digits aligned to their proper places. That is, the number 3,201 differs from 321 because the zero in the tens place ensures that the 3 is understood as being in the thousands place and the 2 in the hundreds, where without the zero, they become, respectively, hundreds and tens.

Further, if we exchange the digits with each other in a positional system, we get a different number. This is quite unlike an additive system, where order does not alter value. For example, if we rearrange 1,231 as 2,311, the four digits are multiplied by different quantities than they were in the previous order:

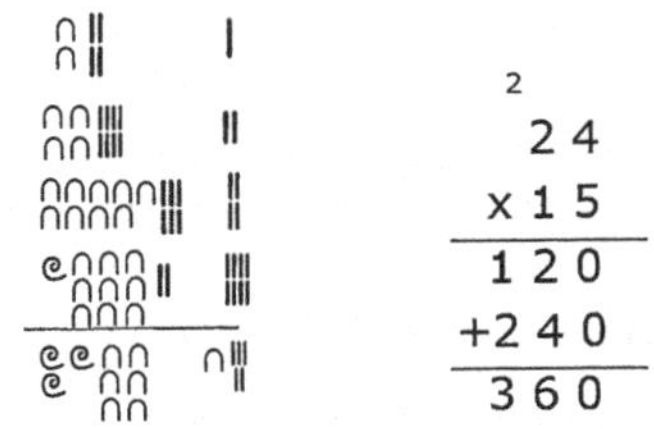

Fig. 3.4 The multiplication 24 × 15 = 360 in (left) the Egyptian and (right) Western methods. Image by the author

$$2\times10^3+3\times10^2+1\times10^1+1\times10^0=2\times1{,}000+3\times100+1\times10+1\times1$$
$$=2{,}000+300+10+1=2{,}311$$

An ancient Egyptian scribe would have found it difficult to understand why 1,231 is so different from 2,311 because he was accustomed to an additive system, where symbols always have the same numerical value even if their places are swapped, and thus, their sum always represents the same value. (Remember: Egyptian symbols are not multiplied by any term, but only added, in an additive system.)

If we compare the two types, the positional system requires few symbols and therefore less space to represent the same numbers and calculate with them, and this is one of the reasons why the Western system has become so widely used. To verify this, write the number 9,999 with the Egyptian system. Instead of the four digits needed in our positional system, you will have to use 36 digits. Or, consider the multiplication 24 × 15 = 360 (Fig. 3.4). With the Egyptian additive system and its associated calculation method, 75 symbols are needed. In comparison, with the Western positional system and its multiplication algorithm, only 14 digits are required.

Calculations become voluminous when additive number systems are used. The positional numeral system with zero allows calculations to be carried out faster and with fewer symbols. It is a more complex system compared to the additive one because it requires the user to understand place value.

In the next chapter, we will return to the numerical quipu to understand whether the decimal number system it uses is positional (with a zero) or additive (without a zero).Competing InterestsThe author has no conflicts of interest to declare that are relevant to the content of this chapter.

Chapter 4
Do Quipus Represent an Additive or Positional Number System?

Abstract In this chapter, we learn there is no certainty that a positional numerical system is present on the quipus and that the answer can only come from the identification of the calculation algorithms used by the Incas.

Keywords Quipu · Decipherment · Additive numbers · Positional numbers · Place value

Locke (1912) classified the numerical system he deciphered on quipus as a decimal system. While this definition is correct, it is incomplete. It is like identifying a person by only his first name, and then not saying which "John" of several is meant. In addition to saying "decimal," Locke should also have specified whether the numerical system was additive or positional. Since Locke, scholars have invariably read the numerical system on quipus as a positional decimal system.

If, on the other hand, we ask ourselves the question, "Is Locke's number system on quipus positional or additive?" and we analyze the structure of the quipu in order to be able to respond, we realize that we reach a stalemate. In fact, both number systems work on the quipu, and the numbers on them can be read in either way, always yielding the same value. There is only one quipu form, with the same knots in the same positions, but it can be read differently, depending on the reader's predisposition to use one number system or the other. It is the classic case of the "half-filled glass," where the optimist sees a glass half full and the pessimist sees one half empty. The glass is the same, with the same amount of water; it is the observer who uses different premises and comes to a different conclusion.

For example, let's consider a string of a quipu on which the number 204 is represented, and analyze the two different visions (Fig. 4.1):

On the left is shown the interpretation of the observer who reads the quipu using positional number system: the reader considers the string divided into sectors at different heights, like a building formed by floors, one above the other. In the highest sector, that of the hundreds, there are two knots; the observer identifies the two knots as a single entity that indicates the number 2. The number 2 must then be multiplied by 100 (10^2), which yields the value 200. There are no knots in the

C. Florio, *The Mathematical Enigma of Guamán Poma's Yupana*,
https://doi.org/10.1007/978-3-031-92662-4_4

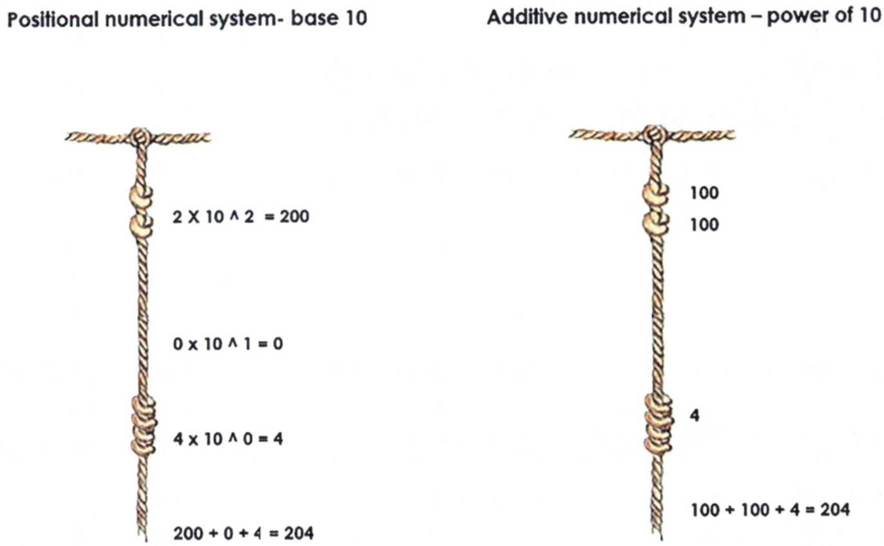

Fig. 4.1 The number 204 on a quipu as interpreted positionally (left) or additively (right). (Image by the author)

position of the tens; having established that the absence of knots indicates the number zero, the number zero is multiplied by 10 (10^1), which yields zero. Finally, the Franciscan knot with four turns indicates the number 4 in the units position, which must be multiplied by 1 (10^0) to obtain the value 4. The result is read as follows:

$$\left(2\times10^2\right)+\left(0\times10^1\right)+\left(4\times10^0\right)=\left(2\times100\right)+\left(0\times10\right)+\left(4\times1\right)=200+0+4=204$$

On the right, we see the interpretation of the observer who reads the quipu using additive number system. As before, the observer considers the string as divided into sectors at different heights, like a building formed by floors one above the other. However, he considers that on each floor, all the knots present are separate but equal entities; further, he sees the value of each knot as changing according to the floor on which it is located, giving it a value of 1, 10, 100, etc. Consequently, the value of a knot is determined by the area in which it is tied, and it is not multiplied by a coefficient (this was also true of the Egyptian symbols, which were not multiplied by a coefficient, but only counted one after another). So, the observer sees two knots in the zone of the hundreds, and for him, each knot corresponds to 100 because all the knots in that zone have the value 100. In the area of the tens, he does not see any knots, so he proceeds as there is nothing to count. Finally, he reaches the units area, where he finds a Franciscan knot with four turns and which therefore indicates the value 4. Adding up the values of all the knots gives 100 + 100 + 4 = 204.

The two strings are identical, with the same knots in the same positions, and the same number is read, but it can be reached using two different and above all conceptually very different numerical systems. At this point, we must wonder how we might understand which of the two numerical systems the Incas really used.

The fact that both number systems can coexist on quipus, although obviously only one can be the correct one, is due to two factors. The first factor was the choice by the Incas to use a string to represent numbers. This meant that knots were used as numerical symbols, as a string can only be knotted.

Let's start our analysis by assuming that the Incas used an additive system: wanting to differentiate the knots indicating units from those meaning tens or hundreds, they had only two possibilities: they could differentiate the knots by either their shape or their position on the string. With string, no other variables are really possible.

In the first case, the use of knots of different shapes to represent the decimal powers—in the same way the units are differentiated by a different knot than used for any higher power—would have immediately marked the presence of an additive system, as the values for the tens, hundreds, and so on would have depended on knot shape, rather than knot position; Of course, the knots would most likely have been arranged in ascending or descending order, to permit easier reading, as was the case for Egyptian numerals. We then would have found ourselves in front of a quipu of the kind hypothesized by the Prince of San Severo (Di Sangro Sansevero 1750) seen in Chap. 2.

On the other hand, making knots of various shapes means tying knots that are much more complex than the simple knot found in the quipu, and complex knots take longer to tie. It was therefore more efficient to use knot position to differentiate their decimal values; this allowed the use of only simple knots for the tens and higher, with their value determined by their respective height on the string.

However, the use of position created another problem: how to identify the area from which counting should be initiated. The Inca apparently decided to highlight the units by using unique knots (Solomon's knot for the number 1; Franciscan knots with 2–9 twists to indicate the numbers 2–9). All of this led to the use of the variable "position" of the knot on the string—not because the numerical system was positional, but only because it was the easiest variable to use.

But since the positional variable in our positional number system is strategic and fundamental and our minds have been strongly conditioned by it, we are misled by our familiarity and the way we have been trained to think into perceiving a parallel between our positional number system and the quipus that does not actually exist. Just as we see our numbers as differently valued according their place, from right to left, we see the knots of a quipu as valued in the same way, from bottom to top. Simply, we see in the quipu the same kind of number system as our own.

The second factor that allows both numerical systems to coexist on the quipu is due to the choice of the same number, 10, to group the quantities. Nonetheless, the use of this number is not surprising. Very distant cultures use the number 10 as a reference in counting because it is the number of fingers of both hands: a common factor of the human species. The interpretation of the quipu as representing an additive system is reasonable, economical, explanatorily sufficient, and poses no difficult questions. It also solves the fundamental problem of the absence of a knot of a particular shape to indicate zero: it had not been elaborated because it was not necessary, since the quipu was an additive system. In additive systems, there is no such thing as zero. The absence of a knot representing zero would simply indicate that zero did not exist for the Incas.

Let's now imagine that the Incas had used a positional system. They would have made one to nine simple knots at the various levels, to indicate the numbers 1–9. These numbers would then have been multiplied by 10, 100, or 1,000, depending on the area of the string where the knots were located. On the lower area, Solomon's knot would indicate 1, and the Franciscan knot in several twists the numbers 2–9. But as a positional system involves the knowledge of zero, the Incas would have needed to indicate it with a knot of a particular shape.

Consistent with the idea of a positional system, scholars have assumed that zero was denoted by the absence of any knots along a section of string. This assumption is unsound, not only because wanting to see something that does not exist forces a certain conclusion, but also because the empty space indicating zero is uneconomical. In fact, knots were always made in areas at the same heights along adjacent strings, so knotting adjacent strings meant continuously comparing them with each other. Moreover, different knots were used for the units. All of this required work and precision. But if zero had been know, with a knot of a particular shape to represent it, tying knots into strings would have been a lot easier, as it would not have been necessary to tie the knots at particular heights, comparing them between adjacent strings. Instead, each string would be knotted independently.

For example, if the number to be recorded was 328, it would be enough to make three simple knots, then leave a small space to indicate a separation, then make two simple knots, a small space and eight simple knots. The last number would necessarily represent the units, and it would not be necessary to use a knot of a different shape to indicate it because if there were other digits they would be inserted lower on the string. In a positional system, the final digit always represents the units. Three sequences of knots would immediately identify a three-digit number, namely 328. On the other hand, 328 could not have been confused with 3280 or 3028 because a hypothetical knot representing zero would be added where necessary. With a positional system, there would have been no need to create the levels on the string that served only to display the "invisible" zero; this would have been wasted effort. It would be enough to make a distinctive knot for zero and insert it where necessary.

Since the units in the quipus were represented with knots of different shapes, the concept of differentiating value by means of shape was known. Had the Incas known zero, it would have made more sense for them to indicate it with a distinctive shape, instead of leaving stretches of string blank and then having to compare them to keep the heights the same.

In trying to remove our preconceptions, the simple objective analysis of the facts supports the additive hypothesis, with the absence of a knot for zero, more than it does the positional hypothesis, which sees unknotted stretches of string as a symbol for zero. On the other hand, any stretches of string without knots could certainly not be eliminated, for this could only be done by cutting the string, losing any knots below the cut.

If the Inca system had been positional with zero, it is very reasonable to think that they would have made a knot to represent it, saving themselves a lot of work in the process, and giving the quipus we observe today a different shape.

It is also interesting to note that in his decipherment, Locke assigned a value to each knot, and he did not indicate the "zero" symbol in the areas of knotless string (see Fig. 2.4, which corresponds to the diagram above in Fig. 4.1, right). Thus, while he does not define the Inca system as positional or additive, his notations imply an additive system, even if he did not state this explicitly.

The idea that zero was unknown to the Incas is also suggested by the vocabulary recorded by Guamán Poma (1615) and (Gonçalez Holguin 1607/1842). Both list the Quechuan numbers from the number "one" to "infinity" ("Pantac huno"), and in both cases, no term for zero is provided. This suggests there was no word in the Quechua language for zero, and this supports the idea that this number system did not contain the concept of zero.

There are other considerations in favor of the additive hypothesis. For example, an additive system is consistent with the level of civilization reached by the Incas; this is by no means meant as a pejorative comment, but only an objective one. Consider that an additive system was also used by the highly mathematical Egyptians, whose advanced level of civilization no one doubts. In the Western tradition, positionality emerged only after thousands of years of effort and refinement, and in fact was imported from Inda, with earlier roots in the traditions of Egypt and Mesopotamia.

A possible objection is that the number system on quipus was positional, just lacking an explicit concept of zero. Many cultures, including the Babylonians and Chinese, have gone through this phase with the consequent problem of the ambiguity it allows. Only the zero that came from Indian culture could solve the problem. For example, consider the Chinese method of multiplying shown in Fig. 4.2:

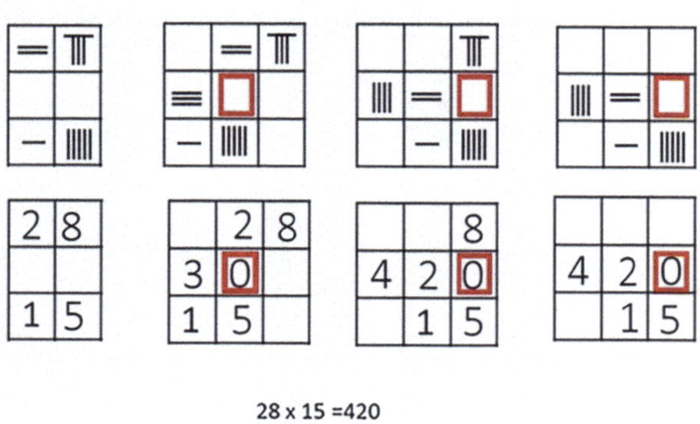

Fig. 4.2 Chinese multiplication of 28 × 15 = 420 using sticks (top) with interpretation in Western numerals (bottom). (Image by the author)

The number system used in the multiplication is positional, but it lacks a zero. This creates the need to arrange the sticks that display the digits on a chessboard to organize them. The absence of any number is not indicated with a symbol, but rather, with an empty square (these empty squares are highlighted in the figure with red rectangles).

To understand whether the numerical system of the Incas was additive for powers of 10, positional in base 10 with the knowledge of zero, or positional in base 10 but without the knowledge of zero, it would be necessary to know the calculation algorithms used by the Incas.

Only by analyzing the calculation algorithms can we understand whether the Incas used zero and whether the numbers were represented with a positional or additive number system, as this information will be evident from the way the numbers are represented and the calculation procedure used.

Knowing the calculation algorithms would lead to unequivocally clarifying the nature of the numerical system used. But how did the Incas carry out their calculations?

Competing Interests The author has no conflicts of interest to declare that are relevant to the content of this chapter.

References

Di Sangro Sansevero, R. (1750). *Lettera apologetica dell'Esercitato Accademico della Crusca contente la difesa del libro intitolato Lettere d'una Peruana per rispetto alla supposizione de' quipu scritta alla Duchessa di S**** e dalla medesima fatta pubblicare [Apologetic letter of the a. Licenza de Superiori.*

Gonçalez Holguin, D. (1607). *Gramatica y arte dveva de la lengva general de todo el Peru, llamada lengua Qquichua, o lengua del Inca.* Francisco del Canto.

Guamán Poma, FA. de. (1615). *El primer nueva corónica y buen gobierno.* Museum Tusculanum Press.

Locke, L. L. (1912). The ancient quipu, a Peruvian knot record. *American Anthropologist, 14*(2), 325–332.

Chapter 5
The Yupana

Abstract In this chapter, we introduce the yupana, the device used by the Incas to perform calculations. We clarify that the yupana does not correspond to the object often found in archaeological excavations to which this name is generally attributed. Rather, only the yupana drawn by Guamán Poma can be considered the one actually used by the Incas.

Keywords Yupana · Guamán Poma · Chronicles

The Chronicles tell us that the Incas calculated with kernels of corn, beans, or colored pebbles; these were manipulated on the ground or a board called *yupana*. The results were then recorded with knots on quipus. One of the Chroniclers, Garcilaso de la Vega, also tells us that if the calculations were simple and involved only addition, subtraction, and multiplication, these operations could be carried out mentally using knots on quipus. It is reasonable, however, to think that the yupana was not only used for division, but also for other arithmetic operations when numbers of a certain magnitude were involved.

As recounted in the Chronicles:

> De la Aritmetica supieron mucho, y por admirable manera, que por ńudos dados en unos hilos de diversos colores dauan cuenta de todo lo que en el reyno del Inca auia de tributos, y contribuciones por cargo y descargo; sumauan restauan, y multiplicauan por aquellos ńudos, y para saber lo que cabía a cada pueblo, hazian las particiones con granos de Mayz y piedrezuelas, de manera que le salia cierra su cuenta." (De la Vega 1609, Book II, Chapter XXVI)

> They knew a great deal of arithmetic and had an admirable method of counting everything in the Inca's kingdom including all taxes and tributes, both paid and due, which they did with knots in strings of different colors. They added, subtracted, and multiplied with these knots, and ascertained the dues of each town by dividing grains of maize and pebbles so that their account was accurate. (as translated by Livermore, 1966, p. 124)

> Pues verles otra suerte de Quipos, q' usan de granos de mayz, es cosa q' encánta. Porq' una cuéta muy embaraçosa, en q' tédrá un muy bué cótador q' hazer por pluma y tinta, para ver a como les cabe entre tát os táto đ cótribució, sacádo táto de aculla, y ańadiédo táto de aca, có otras cié retartalillas, tomará estos Indios sus granos, y ponrnan uno aqui, tres aculla,

C. Florio, *The Mathematical Enigma of Guamán Poma's Yupana*,
https://doi.org/10.1007/978-3-031-92662-4_5

> ocho no se dóde: pasaràn un grano de aqui, trocaràn tres de aculla, y en efecto ellos salen con su cuenta hecha puntualisimamente sin errar un tilde, y mucho mejor se saben ellos poner en cuenta y razon, de lo que cabe a cada uno de pagar, o dar, que sabremos nosotros darselo por pluma y tinta aueriguado." (De Acosta 1590, Book VI, Chapter VIII)

> [To] see them use another type of quipu that employs grains of maize is a fascinating thing. For to make a very difficult calculation, to see how much each person must contribute, which an excellent accountant would have to do with pen and ink, these Indians, taking so many grains from that place and adding a certain number from this, and hesitating a hundred times, will take their grains and put one here, three there, and eight I don't know where; they will move one grain to another place, switch three from elsewhere, and their account comes out very accurately, without the slightest error; and they know much more clearly how to balance an account of what each one has to pay or give, than we could accomplish with pen and ink. (as translated by Martin and Wasserman 2010, pp. 5–6)

> Pero lejos de utilizar solo sus cordeles anudados, los líderes lucanas sacaron otros objetos y los dispusieron en el suelo como parte de una larga y compleja *performance*. Para brindar su testimonio, "leyeron" sus quipus a la vez que trazaban unas rayas en el suelo y distribuían piedritas y granos de maíz en diferentes hileras. (De Anaya 1580–1597, folio 529A)

> But far from using only their knotted strings, the Lucan leaders [i.e, from the province of Lucanas] took out other objects and placed them on the ground as part of a long and complex performance. To bear their testimony, they "read" their quipus while drawing lines on the ground and distributing pebbles and corn kernels in different rows. (as translated by the author)

> Tabla para contar ——— yupana o quippo. (De Santo Thomas, 1560, p. 98)

> Table, for counting ——— yupana, or quipu.

> Los caciques y principales su oficio es holgar y beber, y contar y repartir, que son muy diestros en esto, más que ningún español, y cuéntanlo de espacio y con sus piedras de munchas colores, que cierto es cosa de ver. (De Matienzo 1567, pp. 21–22)

> The chiefs and principles, whose business is to idle and drink, and to count and distribute, are very skilled in this, more than any Spaniard, and they count slowly and with their stones of many colors, which is certainly a thing to see. (translation adapted from Barraza Lescano et al. 2022, p. 226)

Notably, both Father Acosta and De la Vega observed that accountants performed arithmetic tasks in Europe, as if to suggest that they were the only ones in charge of performing such tasks, while also emphasizing the difficulty of the task.

Before dealing with the problem of how the Incas carried out calculations on the yupana, it is necessary to try to clarify what exactly the yupana was, in terms of its shape and size, and the counters used on it. In fact, there are many different opinions about these matters; at least some are mistakes made in the past—errors of simplification in transcriptions and links.

It all began with the 1789 Chronicle by Juan de Velasco. About 200 years after the Conquest, Father Juan de Velasco visited the New World and wrote a chronicle entitled, *Historia del Reino de Quito en la América Meridional*, in which he noted differences between the customs of the inhabitants of the Kingdom of Quito (modern Ecuador) and those of Peru, once the heart of the Inca Empire. The Kingdom of Quito had been one of the last territories conquered by the Incas before they were themselves conquered by Spain. This suggests that the transformation of the Kingdom of Quito into an Inca region (i.e., the adoption of all Inca customs and traditions) had not yet been complete at the time of the Conquest, and this would explain the differences in customs and traditions handed down and described by De Velasco. For the purposes of the present study, the following passage from De Velasco's chronicle is interesting: He notes a difference in how information was stored and how calculations were carried out in the former Kingdom of Quito, compared to method used by the inhabitants of Peru:

> Usaban de una especie de escritura mas imperfecta que la de los quipos peruanos. Se reducía á ciertos archivos ó depósitos hechos de madera, de piedra ó de barro, con diversas separaciones, en las cuales colocaban piedrecillas de distintos tamaños, colores y figuras angulares, ... Con las diversas combinaciones de ellas, perpetuaban sus hechos, y formaban sus cuentas de todo. (De Velasco 1789 [1841], Volume II, Book I, p. 7)

> They [people in Quito] used a kind of writing more imperfect than that of the Peruvian *quipos*. They reduced it [the writing] to certain archives, deposits made of wood, stone, and clay, with divers separations, in which they arranged little stones of distinct sizes, colors, and angular form, ... With the different combination of these they perpetuated their doings and formed the count of all. (as translated in Bingham 1915, p. 186)

In 1869, an interesting object was found in Quito, specifically, in the province of Cuenca. While the original has been lost, there are drawings; these are shown in Figs. 5.1 and 5.2:

Subsequently, many similar specimens or variants were found, though they were far from the Kingdom of Quito and attributed to other pre-Inca cultures. An example is shown in Fig. 5.3:

Various hypotheses have been offered about the meaning and use of these objects. They might have been representations of particular territories or fortresses (Gonzalez Suarez 1892), or they might have been boards for games or counting. Strangely, the chronicles do not mention these fascinating objects, an omission suggesting the Spaniards may have never seen them, perhaps because the Incas kept them hidden or because the Incas did not use them and they belonged to other cultures. The only quote that might refer to objects of this shape is the one from De Velasco given previously, in which he clearly discusses an object from the culture of Quito. However, apart from the first artifact, which was found in Ecuador, the others were

Fig. 5.1 Artifact found in Cuenca, Ecuador, side view (Gonzalez Suarez 1892, pl. III). (Image in the public domain)

Fig. 5.2 Artifact found in Cuenca, Ecuador, bottom view (Gonzalez Suarez 1892, pl. IV). (Image in the public domain)

Fig. 5.3 Pre-Hispanic object from Perù (inv. PAM 1349), probably for ritual use. Federico Balzarotti collection, courtesy of the Municipality of Milan, Mudec Museum of Cultures. (Photo by Studio Neon used by permission. All rights reserved)

found far from Quito (and therefore did not belonging to its culture). But if the object does not belong to the culture of the Kingdom of Quito, it cannot be the object De Velasco described, despite a certain similarity of form. Thus, it cannot be said that it was used to carry out calculations or store information in the Kingdom of Quito or anywhere else. Its meaning and use still remain a mystery.

However, this object is often called yupana and considered the Inca calculating tool. A reasonable assumption that explains this widespread belief is that perhaps in the past the passage from De Velasco was not quoted in full, and as it was copied and translated, the first line was omitted: "They [the people of Quito] used a kind of writing more imperfect than that of the Peruvian *quipos*."

Without the reference to the Kingdom of Quito, readers would tend to believe that the object was used by the Incas, and that in the past, it was associated with the object De Santo Thomas called yupana in his dictionary.

The sequence of events may have been this: an object that did not belong to the Kingdom of Quito was initially mistakenly associated with the words of Juan de Velasco, perhaps because of a similarity in the description and also because the first object of this type was found in Ecuador. The object was therefore supposed not only to store information but also to make calculations in the kingdom of Quito, as De Velasco recounted. Then, because of other finds far from Quito, the object was supposed to belong to other, pre-Inca cultures; it therefore can no longer be linked to De Velasco's description and could have been used for something else. There is no longer any certainty that the object was used in calculations, regardless of the area to which it belonged, but this possible use evidently remains in memory.

Someone simplified De Velasco's passage, omitting the first line and thus eliminating the reference to the inhabitants of the kingdom of Quito; this mistakenly suggested that De Velasco was talking about how the Incas did their calculations.

At this point, someone else presumably made a connection between De Velasco's truncated words (which seemed to refer to the Incas), the object found (which seemed to resemble the object De Velasco described and was thus associated with calculating), and De Santo Thomas' definition of yupana (a counting board). As a result, the object was thought to have been used by the Incas for their calculations, and it was therefore called yupana.

The assembly of a series of approximations and inaccuracies generates false information, but once time has passed and the information has been repeated, it becomes "true."

However, there is no evidence that this object was the yupana used by the Incas for their calculations, or that it was used for calculations in other non-Inca cultures. No known Chronicle or other source of evidence confirms it, but many scholars say it and are convinced of it, perhaps because it has always been said.

To give an example that testifies to the propagation of the error in various publications, consider the following passage from the work of Carlos Radicati di Primeglio, "El sistema contable de los Incas: Yupana y quipu." In talking about the yupana on which the Incas carried out calculations with seeds and stones, Radicati uses De Velasco's quote, without taking into account that the passage explicitly refers to the inhabitants of the kingdom of Quito and not to the Incas, and that it also specifies that the object is used instead of Peruvian quipu.

> En cuanto al instrumento empleado para el cálculo, sólo disponemos de dos fuentes para formarnos una idea de su estructura: La … segunda es la Historia del Reino de Quito, escrita por el padre Juan Velasco, autor tardío pero buen conocedor de las antigüedades indias, que sostiene tratarse de "ciertos archivos o depósitos hechos de madera, de piedra o de barro, con diversas separaciones, en las cuales se colocaban piedrecillas de distintos tamaños, colores y figuras angulares."Es precisamente a base de esta segunda cita que se dio inicio, desde el siglo pasado, a la investigación sobre el ábaco incaico, provocándose al mismo tiempo una polémica acerca del uso de ciertos artefactos hallados frecuentemente en las excavaciones arqueológicas y que, por su gran parecido con los "depósitos" descritos por el padre Velasco, fueron considerados como ábacos, aunque también se los juzgó como tableros de juego e inclusive como maquetas de fortalezas. (Radicati di Primeglio 1979, pp. 270–271)

> As for the instrument used for the calculation, we have only two sources to form an idea of its structure: The ... and the second is the History of the Kingdom of Quito, written by Father Juan Velasco, a late author but a good connoisseur of Indian antiquities, which claims to be "certain archives or deposits made of wood, stone or clay, with various separations, in which pebbles of different sizes, colors and angular figures were placed." It is precisely on the basis of this second quote that research on the Inca abacus began, since the last century, provoking at the same time a controversy about the use of certain artifacts frequently found in archaeological excavations and which, due to their great resemblance to the "deposits" described by Father Velasco, were considered to be abacuses, although they were also judged as game boards and even as models of fortresses. (as translated by the author)

In addition, De Santo Thomas described the yupana as a table, giving the idea of a simple, square and flat object, a physical form very far from the fascinating multilevel object.

The first doubts about the shape of the yupana began to arise in 1908, when a chronicle written around 1615 by Felipe Guamán Poma de Ayala, a descendant of the Inca nobility, was found in the Royal Library of Copenhagen. Entitled *El Primer Nueva Corónica y Buen Gobierno* (or *Nueva Corónica* for short), it was a voluminous chronicle at 1,189 pages, but interestingly, it contained numerous drawings. A facsimile of the chronicle was printed in 1936, and the original was made available on the Internet in 2000 (The Guamán Poma Website). The pages are doubly numbered, as the first numbering was incorrect in several places. In the analysis that follows, references to pages in Guamán Poma's chronicle will give the correct numbering first, followed by the incorrect numbering in round brackets.

Beyond its inestimable value for Americanist studies, this text was a fortunate discovery for scholars interested in Inca numbers and methods of calculation, as page 362 (360) shows a Contador (or *quipucamayoc* in the Quechua language) holding a quipu; the bottom left of the page shows a checkered board with five rows and four columns containing 55 black-and-white counters arranged in a particular repetitive sequence (Fig. 5.4).

In the rightmost column, each square contains a single counter. In the second column from the right, each square holds two counters; in the third column, there are three, and in the fourth (leftmost) column, five. On page 363 (361), which is the reverse of page 362 (360), Guamán Poma describes the Contadores, their duties, and the words for the numbers. In particular, he wrote, "Cuentan en tablas, numiran de cien mil y de dies mil y de ciento y de dies hasta llegar a una." [They count in tables, they number a hundred thousand and ten thousand and a hundred and ten until they reach one.]

The context of the drawing is undoubtedly one of accounting. The table seemed to represent, therefore, the first depiction of a real yupana. Furthermore, it appears even more in line with De Santo Thomas' description. In addition, Guamán Poma also called it "tabla": table. In addition, he wrote "Cuentan en tablas"—they count on tables—using the plural. We will see the importance of this information in the Chap. 9.

Nonetheless, by this point in time, the three-dimensional device described earlier was fairly well entrenched in the literature as the Incan yupana. To differentiate the

Fig. 5.4 Page 362 (360) of Guamán Poma's Chronicle. (Image in the public domain)

two here, the three-dimensional yupana will be called an archaeological yupana, while the second yupana, the checkerboard drawn by Guamán Poma, will be called Guamán Poma's yupana.

Guamán Poma's yupana is seen as a rectangular table with lines that create a fixed grid of five rows and four columns. Radicati di Primeglio saw a confirmation of this rigid structure, in the drawing on a vase shown in Fig. 5.5 (his alternative view will be presented later). He wrote:

> La segunda representación de tablero con escaques la encontramos en una vasija del Museo de Arqueología de Lima que ha sido descrita por L. y Th. Engl (1967: 200, lam.15). La escena de este cántaro consiste en un desfile de personas que transportan con solemnidad un tablero de grandes proporciones, en cuya superficie están deli-neados veinte casilleros (5 × 4), de los cuales la mayoría tienen dos puntos en su extremidad superior. El individuo que carga el tablero estáprecedido por dos guerreros ricamente ataviados y seguido por músicos y por cargadores de trofeos que llevan estacas en cuyas cimas están clavadas cabezas huma-

Fig. 5.5 Drawing by Donna McClelland. Moche archive, Dumbarton Oaks, Trustees for Harvard University, Washington, DC. (Used by permission)

> nas. El dibujo de este tablero es muy semejante, diríamos casi idéntico, al de una viñeta de la difundida crónica de Guamán Poma de Ayala, aquella que ilustra la manera de contar de los antiguos quipucamayos. (Radicati di Primeglio 1979, p. 279)

> The second representation of a chessboard with squares is found on a vase in the Museum of Archaeology of Lima that has been described by L. and Th. Engl (1967: 200, lam. 15). The scene on this pitcher consists of a parade of people solemnly carrying a board of large proportions, on the surface of which are delineated twenty squares (5 × 4), most of which have two points at their upper extremity. The individual carrying the board is preceded by two richly attired warriors and followed by musicians and trophy-bearers carrying stakes on the tops of which are nailed human heads. The drawing of this board is very similar, we would say almost identical, to that of a vignette of the widespread chronicle of Guamán Poma de Ayala, the one that illustrates the way of telling stories of the ancient quipucamayos. (as translated by the author)

Nonetheless, Radicati's idea that the object depicted on the vase is a yupana identical to the one drawn by Guamán Poma is not convincing. The one on the vase, while consisting of squares containing dots, is often found in mochica representations, not necessarily in the 5 × 4 form and with the two dots always at the top of the squares. Instead of a yupana, it is the armor worn by warriors; the armor consisted of square tiles, whose number varied in width and height in order to adapt it to the stature and build of the warrior. The two dots at the top may represent seams or studs that fastened the tiles together, either to each other or to a textile structure beneath them. The presence of armor makes sense, as the procession seems to triumphantly carry the spoils of war in procession: weapons, armor and heads of defeated enemies. This type of armor is represented on many moche pots (Fig. 5.6).

Radicati indicates the resemblance to armor, but the peculiar 5 × 4 structure still leads him to think that it refers to the yupana of Guamán Poma. Later, we will see that Radicati also supports a different vision of the yupana.

Other scholars, besides considering Guamán Poma's yupana to be a fixed squared structure with five rows and four columns, also supposed that Guamán Poma did not draw white counters, but rather, holes dug into the board that served to contain black counters to prevent them from rolling. So, they speculated that the counters in Guamán Poma's drawing were black and that the yupana had the following shape (Fig. 5.7):

To anticipate the different path that we are taking, this study will see the yupana in a different way: as a square table, smooth and without holes, and not fixed with five rows and four columns. Instead, the number of rows and columns will change

Fig. 5.6 Moche bird warrior bottle, fourth–seventh century CE. Gift of Nathan Cummings, 1967 (object number 67.167.1) to the Metropolitan Museum of Art. Public domain

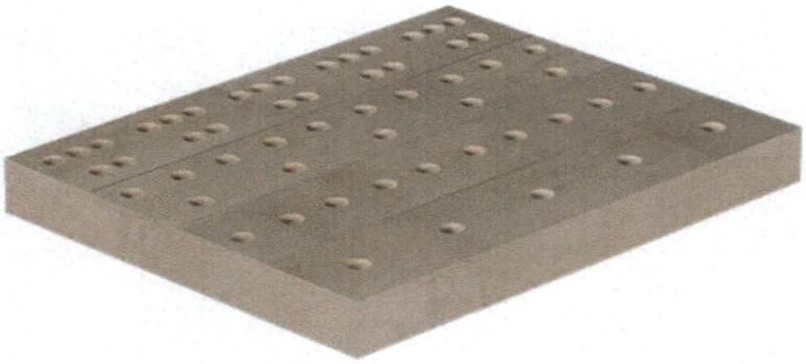

Fig. 5.7 Hypothetical yupana with empty holes. (Diagram by the architect Eliana Continanza and used with permission)

depending on the size of the arithmetic operation to be carried out, in the same way that we, on a sheet of paper, might use several squares for calculating, depending on the space needed for the operation. In other words, the larger the numbers, the more squares we will have to use. Guamán Poma could draw only the part of the yupana necessary for the operation, the one in which the counters were to be inserted, but its size could have been even larger—for example, 20 rows and 10 columns—for a larger calculation.

This study will also show that the *quipucamayoc* may have carried several small yupanas that could be put together to make a larger yupana, as necessary for operations requiring more space. Ultimately, this analysis will conclude that if Guamán Poma had had more space available, he could have drawn the yupana as shown in Fig. 5.8, with many empty squares:

This conclusion may seem surprising at first, but it will later become obvious. The larger form of the 5 × 4 grid was also considered by Radicati, when he

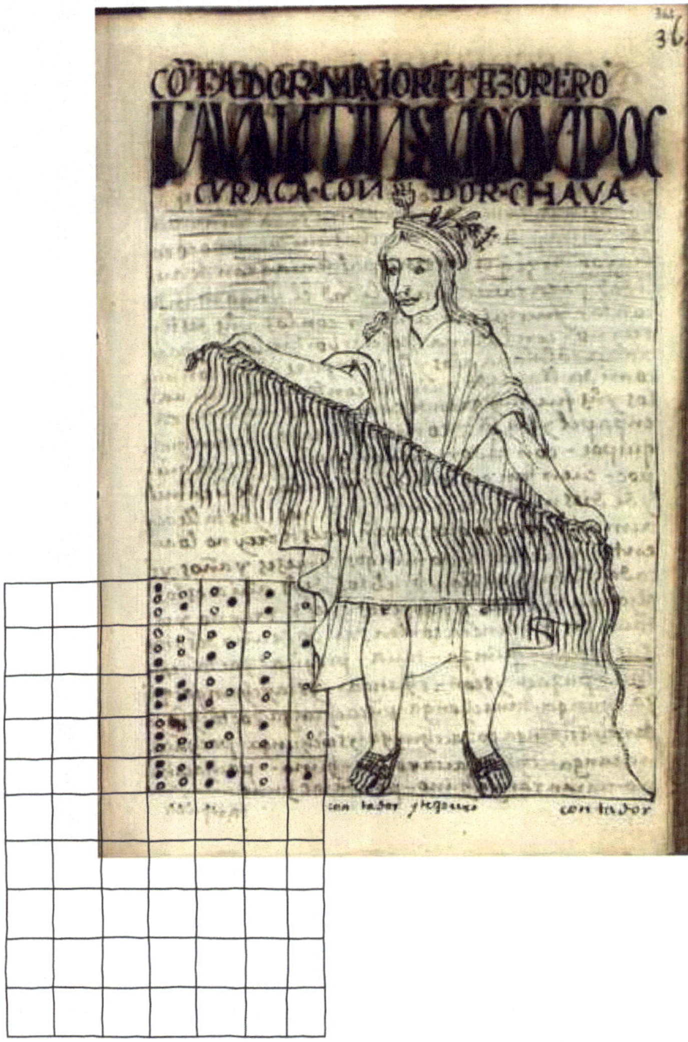

Fig. 5.8 Hypothetical extension of Guamán Poma's yupana for a larger calculation. (Adapted from an image in the public domain)

hypothesized a possible Inca calculation method that we will also analyze later, in Chap. 6.

As for the counters used on the yupana, the Chroniclers clearly speak of corn, beans, and pebbles. Some scholars believe that quinoa seeds were also used, but this is not credible due to the extremely small size of these seeds at 1.5–2.5 mm in diameter. Even the fingertips of a child would have difficulty grasping them to be able to move them on the yupana. This conviction stems from a sentence on page 363 (361) of *Nueva Corónica*, where Guamán Poma wrote of the chief accountant, "Con la lana del cierbo, taruga, enparexaua con la lana a los yndios y enparexaua con una comida llamado quinua [gramínea de altura], contaua la quinua y los yndios. Fue muy grande su auilidad, mejor fuera en papel y tinta."

The sentence is unclear, and this makes it difficult to translate, especially the words "y enparexaua con una comida llamado quinua." Lacking the object complement, it suggests that the quinoa was used for counting and therefore used as a counter. However, the next phrase, "contaua la quinua y los yndios," seems to indicate that the population and the reserves of quinoa were the two variables being counted. In fact, the verb "to count" refers both to the Indians and the quinoa.

It seems reasonable to think that population and food staples were two of the items most counted in the empire. To be precise, quinoa, which had a small size, was counted using a container with a standard volume as a unit of measurement. It was called *ccullu*:

"Per le granaglie era utilizzato il ccullu o phoqca, che equivaleva a circa ½ fanega spagnola, ma che nella stessa Spagna variava di regione in regione e corrispondeva a circa 48 l" (Bonavia 2002). As translated, this means "For grains, ccullu or phoqca was used, which was equivalent to about ½ Spanish fanega, but which in Spain itself varied from region to region and corresponded to about 48 [liters]."

Radicati justified the use of quinoa as a counter by saying its small size was required to be able to insert it into the small compartments of the archaeological yupana. But this hypothesis is not convincing, for two reasons. First, even if the passage were correctly translated as saying that quinoa was used as counters, Guamán Poma wrote about the chessboard yupana, rather than the archaeological one. Second, even if we were to apply the translation to the archaeological yupana, it would still be too difficult to place and extract minute seeds from their small compartments. Doing so would require a pair of tweezers, along with patience and great skill.

Ignoring, for the moment, the number of squares and the presence of holes, the yupana of Guamán Poma consisted of a grid, so a table could also be replaced by lines drawn on the ground. The ground was likely used when counting was conducted before the public, and accordingly, had to be widely visible. Thus, Guamán Poma's Chronicle is the most likely and reasonable evidence for the Inca yupana. As for the counters used, here too it is reasonable to think they were just kernels of corn, beans, and pebbles.

Over time, Guamán Poma's yupana began to interest scholars, who tried to decipher it to understand how the Incas carried out calculations and to discover the algorithms that had once caused De Acosta to write, "es cosa q' encánta" (it is something that enchants) and De Matienzo to say, "que cierto es cosa de ver" (it is certainly a thing to see).

In the next chapters, we will look at some of the most well-known studies that have tried to decipher Guamán Poma's yupana.

Competing Interests The author has no conflicts of interest to declare that are relevant to the content of this chapter.

References

Barraza Lescano, S., Areche Espinola, R., & Marcone Flores, G. (2022). By stones and by knots: The counting and recording of chili peppers stored during the Inca occupation of the Guarco administrative center of Huacones-Vilcahuasi, Lower Canete Valley, Peru. *Andean Past, 13*(1), 221–264.

Bingham, H. (1915). The story of Machu Picchu: The Peruvian expeditions of the National Geographic Society and Yale University. *The National Geographic Magazine, 27*(2), 172–186, 203–218.

Bonavia, D. (2002). I sistemi di misura lineari, ponderali e di capacità: America meridionale. *Istituto Della Enciclopedia Italiana Fondata Da Giovanni Treccani - Riproduzione Riservata.* https://www.treccani.it/enciclopedia/gli-strumenti-dello-scambio-i-sistemi-

De Acosta, J. (1590). *Historia natural moral de las Indias, en que se tratan las cosas notables del cielo, y elementos, mentales, plantas, y animales dellas: y los ritos, y ceremonias, leyes, y gouierno, y guerras de los Indios.* Juan de Leon.

De Anaya, J. M. (1580). *Residencia de Juan Manuel de Anaya, corregidor de la provincia de los Lucanas, Soras y Andamarcas, por Antonio Fernández de Velasco, su sucesor en dicho cargo. Archivo General de Indias, Escribanía 529A [1580–1597].* Ministerio de Cultura, Gobierno de España. https://www.cultura.gob.es/cultura/areas/archivos/mc/ archivos/agi/portada.html

De la Vega, G. (1609). *Commentarios reales, que tratan del origen de los Yncas, reyes que fueron del Peru, De su idolatria, leyes, y gouierno en paz y en guerra: de sus vidas y conquistas, y detodo lo que sue aquel Imperio y su Republica, antes que los Españoles passaran a el.* Pedro Crasbeeck.

De la Vega, G. (1966). *Royal commentaries of the Incas, and general history of Peru. Part one* (H. V Livermore (trans.)). University of Texas Press.

De Matienzo, D. J. (1567). *Gobierno del Perú. Obra escrita en el siglo XVI por el licenciado Don Juan Matienzo.* Compañia Sud-Americana de Billetes de Banco.

De Santo Thomas, D. (1560). *Lexicon, o vocabulario de la lengua general del Peru.* Francisco Fernandez de Cordoua.

De Velasco, G. (1789). *Viaggi, relazioni e memorie relative al regno di Quito.*

Gonzalez Suarez, F. (1892). Lamina tercera. In *Tiempos antiguos ó El Ecuador antes de la Conquista, Segunda Parte, Texto; Atlas arqueologico, Historia general de la Republica del Ecuador* (pp. 67–75). Imprenta del Clero.

Guamán Poma, F. de A. (1615). *El primer nueva corónica y buen gobierno.* Museum Tusculanum Press.

Martin, C. E., & Wasserman, M. (2010). Writing and record-keeping in the ancient Americas: The quipus of the Andes. In *Latin America and its people: Volume 1, To 1830* (2nd ed., pp. 1–13). Prentice-Hall.

Radicati di Primeglio, C. (1979). *El sistema contable de los Incas: Yupana y quipu.* Libreria Studium.

Chapter 6
Chronological List of Studies of Guamán Poma's Yupana

Abstract In this chapter, we report the studies conducted by various scholars on Guamán Poma's yupana, and we begin to understand the issues associated with decipherment.

Keywords Guamán Poma · Yupana

The obvious approach to deciphering Guamán Poma's yupana is to consider it as a type of abacus on which to represent numbers and carry out calculations. The necessary first step is to identify a numerical system that allows numbers to be represented on the yupana, while taking into account the constraints that it imposes: five rows, four columns, white counters, black counters, and the sequence 1, 2, 3, 5. The latter, to some scholars, has suggested the initial portion of the Fibonacci sequence, which would imply that the Incas knew it.

The number of constraints, which is quite high, has comprised a significant obstacle; accordingly, scholars have made various assumptions in an effort to reduce the constraints in order to make a solution more attainable. For example, Henry Wássen 1940 hypothesized that the white circles in Guamán Poma's drawing were not counters at all, but rather, holes dug into the board to accommodate the black counters so they would not roll away. In 1966, Hector Calderón thought that the black counters might represent the value 5, as inspired by the bar in Mayan numbers. In 1971, Emilio Mendizábal Losack assumed that counters of different colors represented numbers that were to be subtracted.

Later scholars, with the exception of Carlos Radicati di Primeglio, and the present author, have followed Wássen's choice to consider the white counters as empty holes and black counters as filled holes. Assuming holes, numbers would be represented by the black counters filling them. They then identify a layout involving four columns, five rows and 55 holes distributed in the sequence 1, 2, 3, 5 as shown in Fig. 6.1. Moscovich (2007, 2016) had a slightly different idea, considering all 55 circles to be holes in two colors—black and white—without the presence of any counters.

C. Florio, *The Mathematical Enigma of Guamán Poma's Yupana*,
https://doi.org/10.1007/978-3-031-92662-4_6

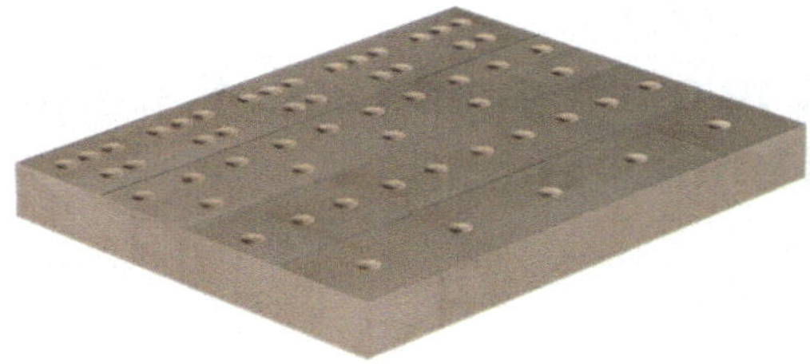

Fig. 6.1 Hypothetical yupana with empty holes, representing Wássen's hypothesis. (Diagram by the architect Eliana Continanza and used with permission)

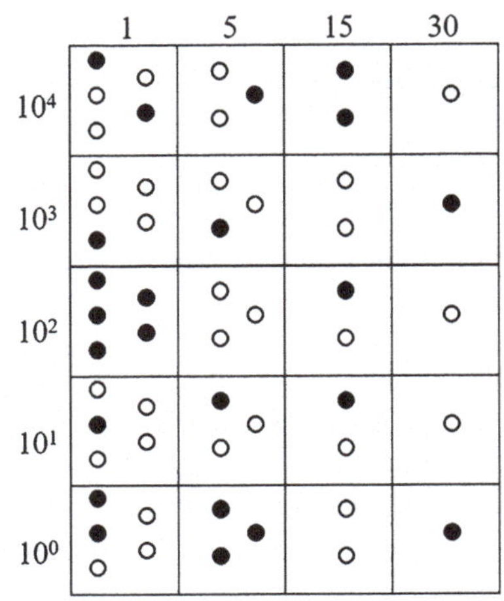

Fig. 6.2 Wássen's interpretation. For example, the counters in the top left square would represent the value 20,000 (2 black counters worth 10,000 each). (Image by the author)

Below are the best-known interpretations of Guamán Poma's yupana as proposed by various scholars, noting that no two agree on all aspects of any interpretation.

Henry Wássen (1931, 1940) hypothesized that the rows (read from bottom to top) represented the powers of 10, while the columns (read from left to right) represented the numbers: 1, 5, 15, and 30 (Fig. 6.2).

Wássen did not explain the number read on the yupana of Guamán Poma; instead, he supposed that the counters were inserted randomly, without wanting to indicate a precise number: "sin intentar marcar un número definido" (without an intent to mark a definite number) (1940, p. 217). However, if the values of all the black counters are added up according to Wássen's scheme, the number 408,257 results.

Carlos Radicati di Primeglio (1979) hypothesized that the columns corresponded to the strings of a quipu, with the rows (read bottom to top) representing the powers of 10 and the counters in all columns having the value 1 (Fig. 6.3). In this case, each counter inserted into a square would corresponded to the value of the power of 10 of that row. In this way, counters of only one color would be used. With no holes on the yupana, the squares would not necessarily contain 1, 2, 3, or 5 counters.

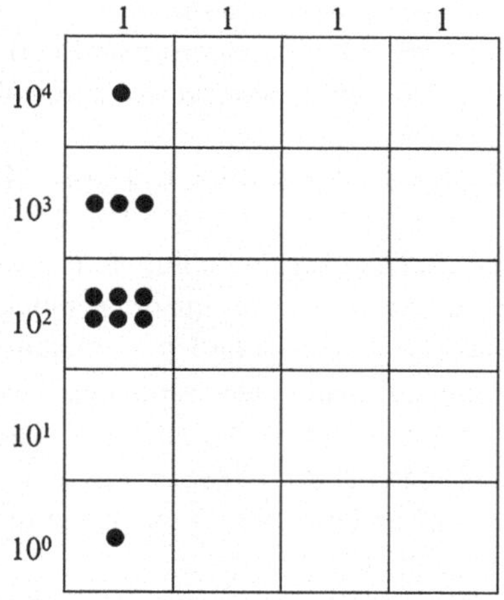

Fig. 6.3 Radicati's interpretation, showing the number 13,601 as an example (as read from bottom to top, ((1 × 1) + (6 × 100) + (3 × 1,000) + (1 × 10,000))). (Image by the author)

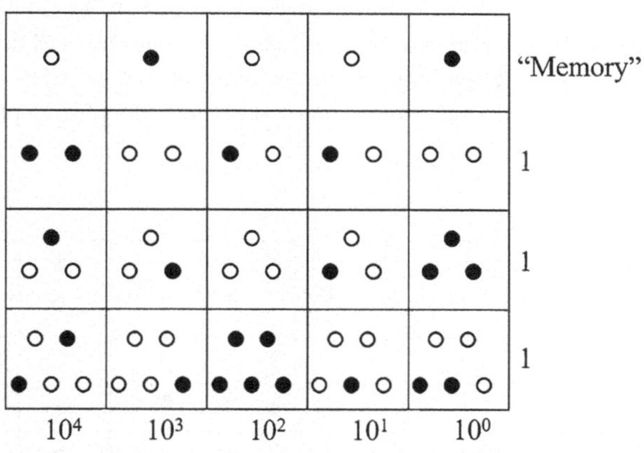

Fig. 6.4 Burns Glynn's interpretation. The rightmost column would represent the value 15, and the column to its immediate left (the column of 10) would represent the value 30. (Image by the author)

Radicati reads four numbers on Guamán Poma's yupana, one for each column counting only the black counters: 21,512; 11,013; 20,110 and 1,001.

As for Guamán Poma's scheme with black-and-white counters in a particular arrangement, Radicati hypothesized that it represented a magical scheme with which numbers with a magical meaning would be obtained through an esoteric procedure.

William Burns Glynn (1981) rotated the yupana 90° to the left (counterclockwise) (Fig. 6.4). Consider the rightmost column: Filling in all the holes of the lowest three rows with black counters would yield the number 10, so these could be emptied and replaced by a single counter in the topmost counter, called "memory," which indicated that there were 10 to be incorporated into the subsequent calculation. The remaining columns would be interpreted identically, except as multiplied by the appropriate power of 10.

If we read the value represented according to Burns Glynn's interpretation, we get the number 62,645.

Nicolino De Pasquale (2001) gave the columns the values 5, 3, 2, and 1 and assumed that the numerical base was 40 or quadragesimal, rather than 10 or decimal. In this case, if the lowest row were completely filled in with black counters, the value 39 would result. Adding 1 would mean replacing these counters with a single counter in the rightmost column of the next higher row, where it would be valued at 40. For Guamán Poma's yupana (Fig. 6.5B), de Pasquale assumed that each row represented a separate number and used 40^0 as the power to read them. He then read the numbers: 17, 9, 27, 10, and 20.

Viviana Moscovich (2007) turned the yupana 180° in order to view it from the perspective of the Contador shown on page 362 (360) of the *Nueva Corónica*, resulting in the interpretation shown in Fig. 6.6. Next to the uppermost left square, as rotated, Moscovich identified a thicker line, speculating that it would indicate the point from which reading was to be initiated. Moscovich saw the 55 black and white circles of Guamán Poma's yupana as 55 black and white holes. These holes were given different values, depending on the square in which they were located. Moscovich speculated that the arrangement reflected the philosophical-religious

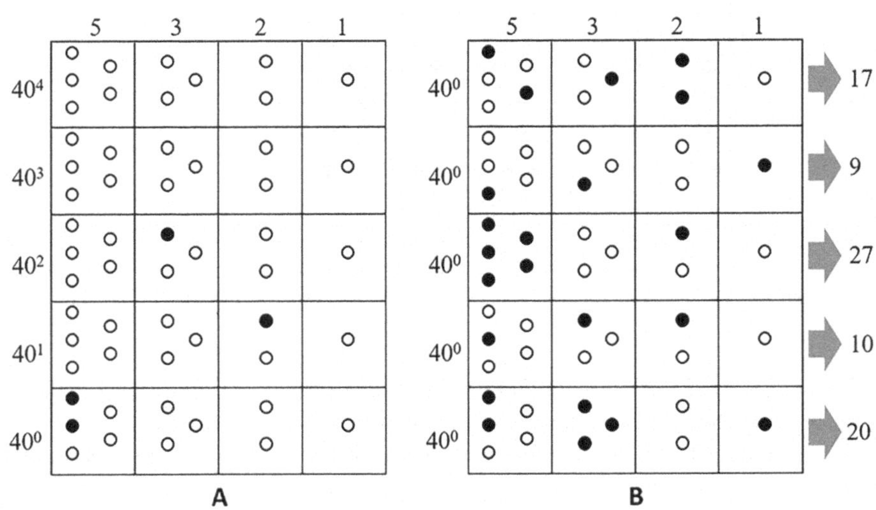

Fig. 6.5 De Pasquale's interpretation. (**A**) The numbering scheme identified by de Pasquale, showing the number 4,890 (that is, $((5 + 5) \times 40^0 + (2 \times 40^1) + (3 \times 40^2))$). (**B**) de Pasquale's interpretation of Guamán Poma's yupana; in this case, de Pasquale assumed the yupana contained five numbers, one per row, each valued at the power 40^0. On this interpretation, the bottom row represents the number 20 (that is, $(((5 + 5) + (3 + 3 + 3) + (1)) \times 40^0)$). (Image by the author)

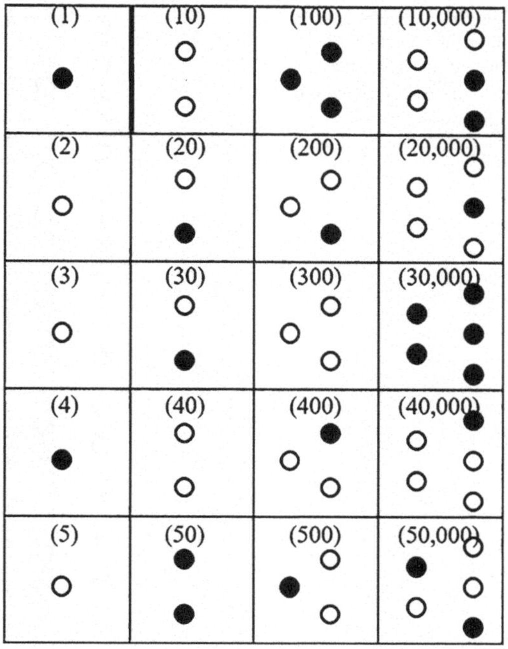

Fig. 6.6 Moscovich's interpretation. In this scheme, the black hole in the top left square, which is worth 1, finds its complement in the white hole in the bottom left square, which is worth 5; similarly, the two white holes at the top of the second column yield the value 10, which is complementary to the two black holes at the bottom of the same column, which are worth 50, and so on. (Image by the author)

concept of Andean duality, "Hanan-Hurin." Her explanation of the values of the holes inside the squares—the reason why they might have been inserted in a sequence that reflected the Hanan-Hurin philosophical concept of duality—is fascinating, but it does not refer to a strictly mathematical logic. Moreover since no counters are used, this interpretation of the yupana on page 362 (360) of the *Nueva Corónica* does not yield any numbers.

Cinzia Florio (2009), the present author, presented a novel interpretation that corresponded exactly to the Incan decimal system and the quipu arrangement. This interpretation is discussed in detail in the Chap. 8, so it is not described here, other than to say that it does not assume a positional number system and finds an algorithm of multiplication.

Andrés Chirinos Rivera (2010) thought that each hole on Guamán Poma's yupana had a value, but he provided two different sets of values. In the first case, he said the yupana reflected numbers in base 66, which gave the holes the following values (Fig. 6.7):

When these values are assigned to the counters on Guamán Poma's yupana, the following number results:

$$11{,}000+7{,}000+5{,}000+3{,}000+2{,}000+900+400+100+110+100+90+80$$
$$+70+30+10+6+3+1.1+1.0+0.4+0.5+0.6+0.1=29{,}902.7.$$

In the second case, Chirinos Rivera assigned the values shown in Fig. 6.8A to the holes on the yupana. Figure 6.8B shows how a number can be represented: the

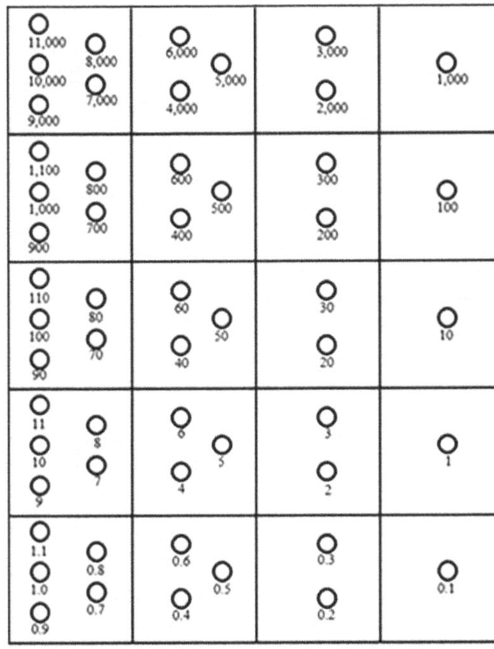

Fig. 6.7 Yupana values according to the first diagram by Chirinos Rivera (2010). (Image by the author)

values of the black balls are added together, but they are only in the positions indicated in the graph.

Applying this scheme to the yupana of Guamán Poma, we obtain the number:

$$(1+1)+(10)+(100+100+100+100)+(1,000+5,000)=6,412$$

Notice that this interpretation requires discarding all the black counters that do not touch the graph, these are circled in Fig. 6.8C.

Of note, Chirinos did not actually apply his method to Guamán Poma's yupana. That is, he did not try to read a number on it, but when this is done, we find the number 29,902.7 with the first diagram and 6,412 with the second.

Molly Leonard and Cheri Shakiban (2011) based their interpretation (Fig. 6.9) on that of William Burns Glynn, with the difference that the yupana was not rotated 90°; there were also some variations in the calculation methods. These authors did not interpret Guamán Poma's drawing, but if it were read according to this interpretation, it would yield the same number found by Burns Glynn (that is, 62,645). Here, too, the column with only one hole in each square is used as "memory" in addition and subtraction operations.

Subhash Kak (2014) hypothesized that the yupana used a positional number system with base 144 (centetetraquadragesimal). Kak assigned the values 1, 6, 24, and 72 to the columns. On this interpretation, if the bottom row were filled in completely with black counters, it would yield the number 143. Adding 1 to get 144 would require replacing these with a single counter in the column with 5 counters

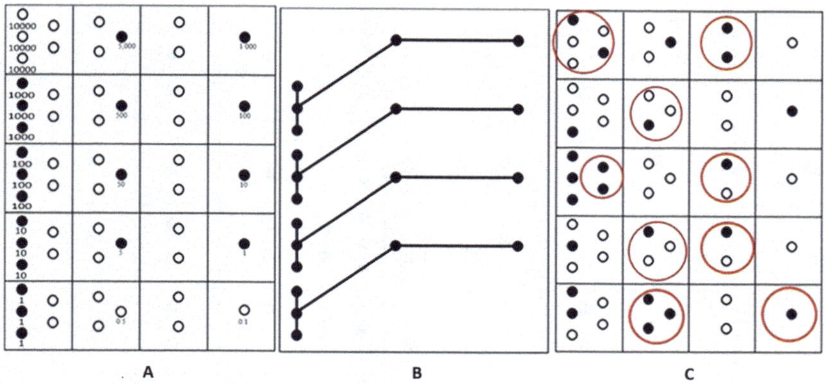

Fig. 6.8 (**A**) The second diagram of Chirinos Rivera with the values given to the counters according to the holes in which they are located; (**B**) the number is represented by the sum of the values of the seeds found only in the holes marked on the graph; (**C**) reading the yupana of Guamán Poma by taking only the black seeds indicated in graph (**B**) and considering only those that are not circled. (Image by the author)

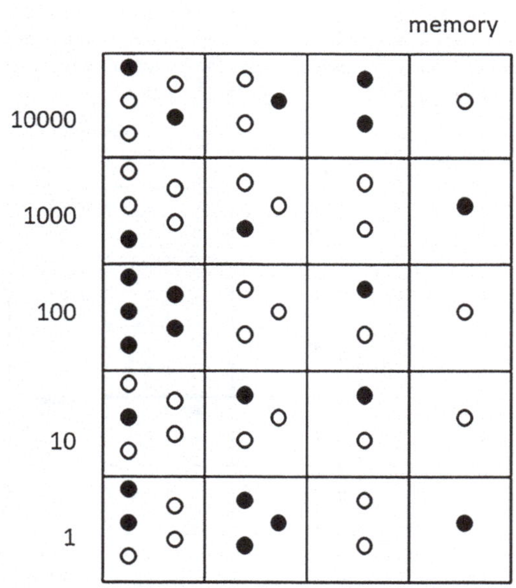

Fig. 6.9 The interpretation of Leonard and Shakiban (2011). (Image by the author)

on the next higher row, where it would be valued at 144. For Guamán Poma's yupana (Fig. 6.10B), Kak assumed that each row would be valued at 144^0. Kak hypothesized that the five numbers read were 56, 79, 29, 31 and 92, purported to be related to astronomical events; the hypothesized connections, however, were not supported by a rigorous demonstration.

Milton Rojas-Gamarra and Marina Stepanova (2015) hypothesized that a horizontal line divided Guamán Poma's yupana into two parts, with rows below the

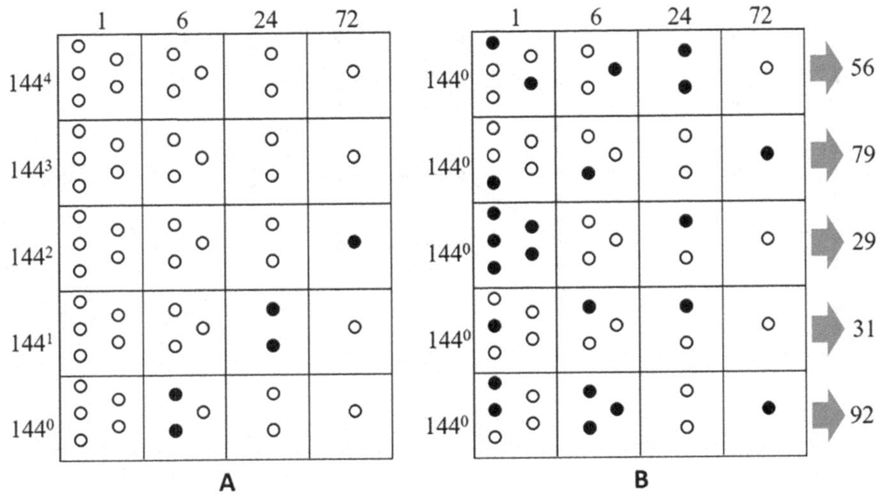

Fig. 6.10 Kak's interpretation. (**A**) The numbering system identified by Kak on Guamán Poma's yupana. In this schema, the figure would show the value 1,499,916 (that is, $((6 \times 2) \times 144^0) + ((24 \times 2) \times 144^1) + ((72 \times 1) \times 144^2)$, which is $((12 \times 1) + (48 \times 144) + (72 \times 20{,}736))$ or $(12 + 6{,}912 + 1{,}492{,}992)$). (**B**) Kak's interpretation of Guamán Poma's yupana; in this case, the author reads five numbers, with the bottom row representing the value 92 (that is, $((1 + 1) + (6 + 6 + 6) + (72)) \times 144^0$). (Image by the author)

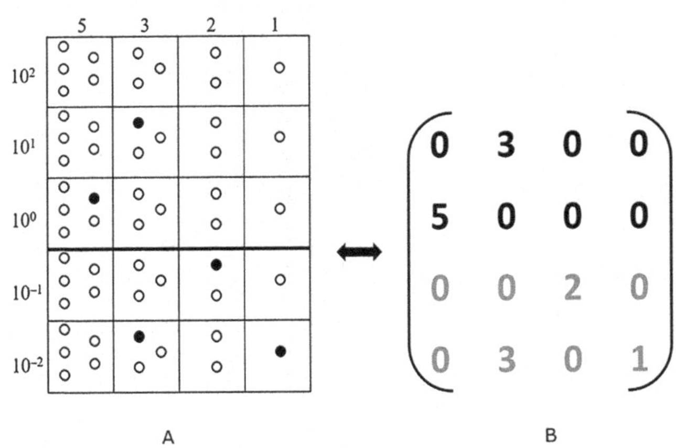

Fig. 6.11 Rojas-Gamarra and Stepanova's interpretation. (**A**) The numbering system identified by Rojas-Gamarra and Stepanova. (**B**) The associated matrix. Both represent the number 35.24. (Image by the author)

line indicating decimal fractions (Fig. 6.11). They gave the columns (read from left to right) values of 5, 3, 2, and 1; they gave the rows the powers of 10, with those above the line assigned positive powers and those below the line negative ones. An associated numerical matrix, when populated with a suitable calculation, would

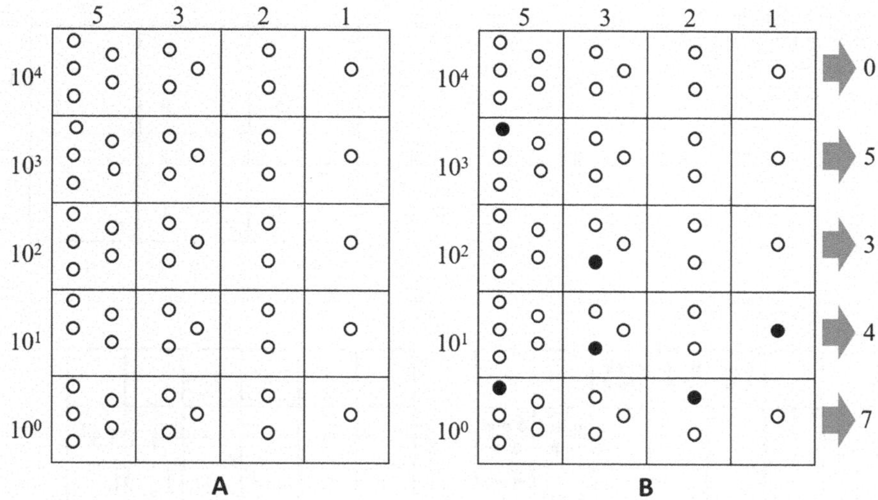

Fig. 6.12 Saldívar Olazo et al.'s interpretation. (**A**) The numbering scheme identified by Saldívar Olazo et al. (**B**) Example: on this scheme, the counters of a single color, when placed in the squares and read from top to bottom, would form the number 5,347 (that is, $(5 \times 10^3) + (3 \times 10^2) + ((3 + 1) \times 10^1) + ((5 + 2) \times 10^0)$ or (5,000 + 300 + 40 + 7)). (Image by the author)

produce the number represented on the yupana. If this reading were applied to Guamán Poma's yupana, the number would read: (1,700 + 90 + 27 + 1 + 0.2) = 1,818.2.

Saldívar Olazo et al. (2019) thought that the rows (read from bottom to top) had the values of the powers of 10 in increasing order, and that the columns (read from left to right) had the values 5, 3, 2, 1 (Fig. 6.12). Applying this method to Guamán Poma's yupana yields the number: (170,000 + 9,000 + 2,700 + 100 + 20) = 181,820.

After proposing various ways in which numbers could be represented on Guamán Poma's yupana, all these scholars also developed various algorithms for conducting arithmetic operations on it; these algorithms differ for each scholar. Ultimately, anyone can write numbers on the yupana, perform arithmetic operations on it, and receive consistent results, no matter how many different ways are used. For the sake of brevity, the various algorithms of each scholar are not reported here. Only one example will be shown: the multiplication according to the hypothesis of Carlos Radicati di Primeglio.

Radicati was inspired by the calculus hypothesis developed by Calderón (1966) for the Mayan culture, and with appropriate variations he elaborated the following method which, it must be said, is very similar to the matrix multiplication used in Europe after the arrival of the positional system from the East (Fig. 6.13). It is also called *gelosia* (literally, "jealousy") multiplication, a name derived from the matrix's graphical resemblance to the grates on Venetian windows that once shielded the ladies within from prying eyes.

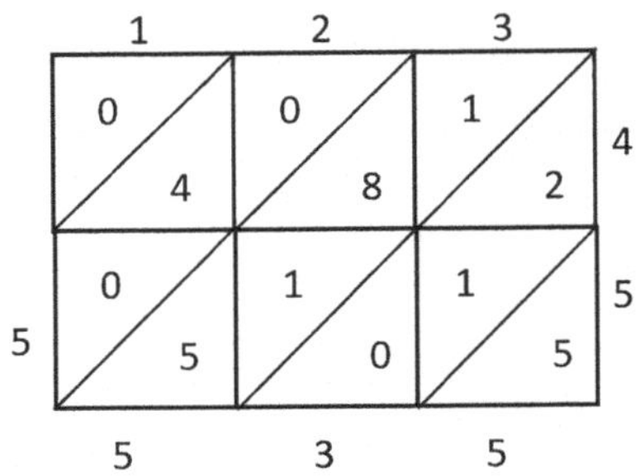

Fig. 6.13 Example of matrix multiplication: 123 × 45 = 5,535. (Image by the author)

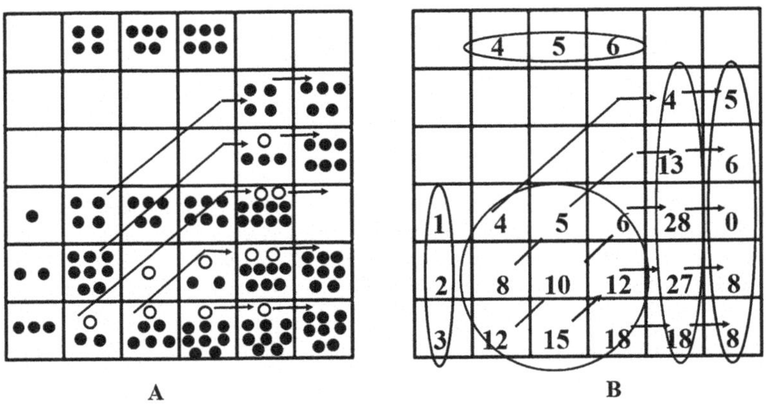

Fig. 6.14 The multiplication 123 × 456 = 56,088 using the Radicati method, (**A**) with the counters used by the Incas; (**B**) in Western numerals. (Image by the author)

Figure 6.14 shows the multiplication 123 × 456 using the Radicati method. Each digit of the first number in the figure is multiplied by each digit of the second (the numbers inside the smaller ovals). This yields the numbers in the circle. Then the sums are carried out diagonally, obtaining the numbers in the first large oval; finally, starting from the bottom and going to the top of the first large oval, the second digit of the number encountered is transcribed in the second large oval furthest to the right, while the first digit is added to the number of the upper line, and so on. The result obtained (which is in the large oval furthest to the right) is 56,088. Radicati's interpretation will discussed later to make an interesting observation about the studies of this passionate scholar.

We therefore find ourselves with many possible interpretations—that is, with many possible hypotheses—none of which, however, can be proved. To many readers, this concept may not be clear, as they may think that if numbers can be written and read on the yupana and calculations performed with them, it must mean that the yupana has been correctly deciphered. But since each scholar gives a different version, readers should be asking the question: what were the actual numbers and algorithms used by the Incas? How do we know who is correct, and why are there so many potential answers, only one of which is likely to be right?

In translating a linguistic message in an ancient language, different scholars do not usually read completely different things. Rather, the language either remains undeciphered, or it is deciphered in such a way that everyone generally agrees that the actual text has been recovered. Reader should therefore begin to understand that deciphering a mathematical message is different from deciphering a linguistic one.

Now the full Pandora's box of questions must be opened: Why is it possible to read such different numbers on the yupana and to perform such different calculations on it? Why is it the case that deciphering a stele in an ancient language will yield a single translation, but a numerical one has an infinity of possibilities? If anyone can come up with yet another possible interpretation, how do we know who might be correct, and whether any particular interpretation is the right one? Why is it possible to identify so many different number systems and so many different algorithms on the yupana? What is the decisive factor, the proof that we must have to understand that we have really deciphered the yupana of Guamán Poma?

To answer these questions, we must first divide the problem into two parts. First, we must answer the questions of why it is possible to read different numbers in different number systems on Guamán Poma's yupana and how we can tell if a scholar has identified the numbers that Guamán Poma really wanted to represent. Second, we must answer the question why everyone is able to make calculations on yupana with consistent and identical results, but using completely different methods, and of course whether any one of these methods was the one actually used by the Incas.

To a mathematician, the answers to these questions will be trivial, so we must ask their pardon for the upcoming discursions into material intended to help specialists in other fields, in which the subject will be treated in terms that are not strictly mathematical. Nonetheless, we hope that this will allow all readers to approach the questions and issues raised and understand the answers. Of course, those who are mathematically inclined can always skip it.

Competing Interests The author has no conflicts of interest to declare that are relevant to the content of this chapter.

References

Burns Glynn, W. (1981). La tabla de cálculo de los Incas. *Boletin de Lima, 11*(3), 1–15.

Calderón, H. M. (1966). *La ciencia matematica de los Mayas*. Editorial Orión.

Chirinos Rivera, A. (2010). *Quipus del Tahuantinsuyo: Curacas, Incas y su saber matemático en el siglo XVI*. Editorial Commentarios.

De Pasquale, N. (2001). Il volo del condor. *Rivista Dell'Ordine Degli Ingegneri Della Provincia Di Pescara*, 0, 1–19. http://www.quipus.it/DePasquale_Volo_del_Condor.pdf

Florio, C. (2009). Encuentros y desencuentros nella individuazione di una relazione matematica nella yupana in Guamán Poma de Ayala. In E. Guagliano (Ed.), *Incontri e disincontri tra Europa e America, Atti del XXX Convegno Internazionale di Americanistica Salerno, 14–15 maggio e 10–12 dicembre 2008* (pp. 151–186). Oèdipus Editore.

Kak, S. (2014). Guamán Poma's yupana and Inca astronomy. Preprint. In *arXiv*: Vol. 1401.7637. https://arxiv.org/ftp/arxiv/papers/1401/1401.7637.pdf

Leonard, M., & Shakiban, C. (2011). The Incan abacus: A curious counting device. *Journal of Mathematics and Culture*, *5*(2), 81–106.

Mendizábal Losack, E. (1971). *Estructura y función en la cultura andina. Inka phase* [Unpublished doctoral dissertation]. Universidad Nácional Mayor de San Marco.

Moscovich, V. (2007). Yupana, tabla de contar inca: Estructura interna. *Revista Andina*, *44*, 71–116.

Moscovich, V. (2016). *El khipu y la yupana: Administración y contabilidad en el Imperio Inca [The khipu and the yupana: Administration and accounting in the Inca Empire]*. Ediciones El Lector.

Radicati di Primeglio, C. (1979). *El sistema contable de los Incas: Yupana y quipu*. Libreria Studium.

Rojas-Gamarra, M., & Stepanova, M. (2015). Sistema de numeración Inka en la yupana y el khipu. *Revista Latinoamericana de Etnomatemática: Perspectivas Socioculturales de La Educación Matemática*, *8*(3), 46–68.

Saldívar Olazo, C. G., Saldívar Olazo, A. J., & Goycochea Olazo, D. (2019). Tawa pukllay – La aritmética Inca de reconocimiento de formas y movimientos operable en paralelo y que no requiere cálculos numéricos mentales. *Acta Latinoamericana de Matemática Educativa*, *32*(1), 354–363.

Wassén, H. (1931). The ancient Peruvian abacus. *Comparative Ethnographical Studies*, *9*, 189–205.

Wassén, H. (1940). El antiguo ábaco peruano según el manuscrito de Guamán Poma. *Ethnologiska Studier*, *11*, 1–30.

Chapter 7
Differences Between Deciphering a Linguistic Message and a Mathematical One

Abstract In this chapter, we offer logical ideas for understanding the necessary and sufficient requirements of decipherment, in order to be sure of having managed to decipher a mathematical message in an unknown language.

Keyword Mathematical deciphering

To explain the differences between deciphering a linguistic message or a mathematical one, I will offer a premise; for greater clarity, I will use an example.

Consider a language—for example, English—and imagine that we have a box that has the particularity of gradually increasing its volume, as we fill it. We put in this box, one at a time, all the words that belong to the English language. The volume of the box will increase, it will become very large, and when we have finished entering all the words that belong to the English language, it will stop at a very large but very precise value. Mathematicians would call the box a "finite set."

Why do words run out at some point?

Words are "assembled" with the 26 letters of the alphabet, so the combinations could be infinite, but in a language there are two limits: the first is that words are made up of a limited number of letters, mostly from 1 to 10 letters. There are longer words and very long words, but there is a maximum number of letters used. The longest word in English is 45 letters long: pneumonoultramicroscopicsilicovolcanoconiosis (it is a type of lung disease).

The second is that if the letters that make up a word, we recombine them in a different way, we will not necessarily realize words that exist in that particular language. A word doesn't always have an anagram. For example, the word "mom" exists, but the words "mmo" and "omm" do not exist; and if a word has an anagram, it is difficult for it to have more than one. So even if we have 26 letters that can be repeated several times, we use a limited number of letters to compose a word, and we don't assemble them in all possible combinations. All this means that there are many words, but of a finite quantity.

Now we use the same box, but this time, instead of all the words in the English language, we enter all the numbers we can write with our number system. We would

C. Florio, *The Mathematical Enigma of Guamán Poma's Yupana*,
https://doi.org/10.1007/978-3-031-92662-4_7

soon realize that we would never finish filling the box, as there would always be another larger number to enter. The box would become very large, but above all endless... infinite. We would find ourselves in the presence of an infinite whole.

Why is it that unlike words in a language that are limited, numbers are unlimited? It's because we don't put the same limits on numbers as we do on words. In a word, we limit the number of letters, but a number can be 10 digits, 100 digits, and we can still reach the number of digits we want. For this reason, the need arose to create a notation, called scientific notation, that allows us to write a very large number in a compact form. For example, the number 15700000000000000000000000000000 can also be written as 1.57×10^{31}.

As has already been said, in words, letters cannot be changed from place to create a word that exists, not all words have anagrams, and in any case it is difficult for them to have more than one. On the other hand, in each number, we can change the place of the digits at will, and we will always get a number that really exists. For example, if we consider the number 123, by changing the position of the digits, we can have five other different but meaningful combinations:

132 213 231 312 321.

All these numbers also exist, precisely because the numbers are infinite. Using "incorrect" terminology, we could say that every "anagram" of a number is a number too.

In other words, if with my eyes closed I take random letters and put them next to each other, it is not certain I will compose a real word. Instead, it is very likely that by opening my eyes, I will find myself pronouncing a strange sequence of sounds, like a child who does not yet know how to speak. However, if instead of letters I use digits, when I open my eyes I will surely be able to read a number that exists, as all possible arrangements of digits will always give me an existing number. The only problem I would have if I found one or more zeros placed at the beginning—for example, if I shuffled the digits 3, 7, 0, 1 with my eyes closed—I could find them positioned like this: 0713. In this case, I would read the number 713, discarding the zero, which would not make sense.

Having said this, let's try to consider the decipherment of a linguistic message; let's start with a sentence with a complete meaning, with a subject, a verbal predicate and complements in which about 55 letters are used, parallel to the number of elements present in the yupana of Guamán Poma (1615) that we will try to decipher later (55 circles):

Analysis lessons today will be held in the computer science department.

Let's suppose we encrypt this message, and since we are not experts in encryption, we do it in the most trivial way: that is, we associate each letter of the alphabet with a particular symbol chosen at random (Fig. 7.1):

Then, we replace all the letters with symbols. We will get the following unreadable message (Fig. 7.2):

a	b	c	d	e	f	g	h	i	j	k	l	m	n	o	p	q	r	s	t	u	v	w	x	y	z
!	$	&	=	^	:	%	/	(	@	;	“	\	£	#	°	.	~	)	,	]	●	+	[	■	-

Fig. 7.1 Random associations between letters and symbols. (Image by the author)

!£!“■)() “^))#£) ,#=!■ +(“ “ $^ /^“= (£ ,/^ &#\°],^~)&(^£&^ =^°!~,\^£,

Fig. 7.2 Unreadable message. (Image by the author)

The symbols will remain meaningless, unless we know the key to interpreting it: the correspondence letters of the alphabet to the cryptic symbols shown above.

Then we ask another person, who is also not an expert in decoding, to decrypt the encrypted message. The only information we provide him with is this:

1. The message is a linguistic message, that is, it contains only words made up of letters; there are no numerical digits.
2. The message is written in the English language.

Our friend, not being an expert, will move in the most trivial way, so he will start by randomly linking each symbol of the encrypted message to a letter of the English alphabet, and then he will see if he will be able to read a message of complete meaning. We all agree that, unless he has boundless luck, our friend will certainly not succeed in deciphering on the first attempt, and certainly not even in the next one or in many others. How will he know he didn't succeed? The question may seem trivial because we move in the linguistic field, and therefore we can all answer, but we do it anyway in order to be able to better understand the parallel that will be made later in the mathematical field.

He will realize that he has made a mistake because when he goes to read—or rather, tries to read—the attempt to decipher, he will find himself in front of words that do not exist in the English language. They will only be sequences of letters that are difficult to pronounce and meaningless: in the end, he will not be able to read them. Then he will continue to replace symbols with letters, again changing correspondences, hoping to be able to read known words.

In the various attempts, he may also be able to identify, in the midst of various nonsense words, some words belonging to the English language, especially if they are composed of just a few letters, for the formation of anagrams. However, he will never read all words that belong to the English language, that are all well-formed, and that nonetheless differ from the original message. It is evident that such a case is impossible on such a long sentence; the existence of many anagrams would be necessary. Ultimately, either he can read the sentence we had encrypted or he always reads nonsense.

So our friend will keep trying until he reads the sentence:

Analysis lessons today will be held in the computer science department.

"Analysis lessons today will be held in the computir science department"

Fig. 7.3 Example of mistake. (Image by the author)

At this point, he will do something trivial but important: he will stop and say that he has deciphered the message. Because he can read a sentence, he stops; he doesn't even consider the possibility that there might be another decipherment. On such a long sentence, with the limits that a language imposes on the number of possible words as well as the correct grammatical construction, the possibility should not even be considered. Not only that, he would stop and say that he had deciphered the sentence, even if there was a mistake and a letter was wrong. He would say that there was a typo, and all of us would broadly agree with him. Example (Fig. 7.3):

If, on the other hand, our friend had stopped at the first attempt at decipherment and thought it was correct, we would have been stunned and asked what he understood about all those strange sounds. No matter what explanation he tried to give us, he wouldn't have convinced us, and we would have thought that our friend had suddenly gone mad.

All this reasoning, which appears clear in the linguistic field, we will exploit in a parallel in the mathematical field in order to better understand what happens in this other sector.

So this time let's consider a mathematical message. What's changes? Instead of letters, we will use digits; instead of words that are obtained by assembling letters, we will have numbers that are obtained by assembling digits; and instead of a proposition that binds words together to obtain a meaningful message, we will have a calculation algorithm that logically binds numbers to obtain a result. Let's consider the addition algorithm, which is the simplest, most intuitive and well-known, whose components are the addends and the sum. These parallel the subject, the predicate and the complements in the linguistic field. A mathematical message of this kind is often delivered to us, for example, every time we go to a supermarket to do our shopping: the receipt.

On the receipt, there are a series of numbers that have no relation to each other—the costs of the various products—but at the end, the sum appears, and immediately all the numbers become part of a logical message. It will be a mathematical message and not a linguistic one, but a message nonetheless. It will tell us the cost of the various products and the consequent amount of money we have to pay at the checkout.

So let's take for example the following mathematical message in which we used 18 numbers with 55 digits in total (again, to stay within the scope of the 55 symbols used on the yupana of Guamán Poma):

23,15,470,97,345,22,987,121,223,453,1111,7786,8888,8953,245,643,87,**30469**

The last number is the sum of all previous numbers.

In parallel to what we did with the linguistic message, let's encrypt it by linking each digit to a randomly chosen symbol (Fig. 7.4):

0	1	2	3	4	5	6	7	8	9
£	)	$	\	/	%	(	?	!	$

Fig. 7.4 Associations linking digits and symbols. (Image by the author)

$\ )% /?£ $? \/% $$ $!?)$) $$\ /%\ )))) ??!(!!!! !$%\ $/% (/\ !? \£/($

Fig. 7.5 Encrypted message. (Image by the author)

By substituting symbols for digits we will have the following encrypted message (Fig. 7.5):

We summon the usual friend and ask him to decrypt the encrypted message by providing him with the following two pieces of information that were also given in the corresponding linguistic scope:

1. The message is mathematical, so there are only numbers and not words.
2. The number system used is our base-10 positional system with the digits 0–9.

The decipher here too will randomly connect the symbols to the digits 0–9 and replace them. Again, unless he's very lucky, he'll make a mistake. He will also make a mistake the second time, and many more. How will he know he's wrong?

There is a difference here with respect to the linguistic scope. In fact, before, in the linguistic message, we realized we were wrong because the deciphered words did not exist in the English language; we only got strange sounds that in practice, we could not read; Not being able to read gave us the immediate understanding that we had made a mistake in deciphering.

In the mathematical field, on the other hand, when symbols and digits are replaced, even on our first attempt, we get numbers that can be read without any difficulty. This is because numbers are infinite, and every combination of digits is a number that exists, as we saw earlier with the example of the box.

Being able to read numbers, however, can naively lead us to think, already on the first attempt, that we have deciphered the message, as being able to read something gives the feeling of having succeeded in the intent, conditioned by the parallelism with the decipherment of a linguistic text where, if we can read words instead of random sounds, then we have succeeded.

So the difference between deciphering a linguistic message and a mathematical one is that for the first it happens that, either we decipher the sentence or we always read non-existent words, while for the second we always read sequences of existing numbers, but only one sequence is the right one and we can't figure out which one it is, since we can read them all.

Ultimately, in the linguistic field, if we can read words, we have certainly deciphered the message, but in the mathematical field, if we can read numbers, it is not at all certain that we have deciphered the message because we would still read numbers.

This is why each scholar can read different numbers on the yupana: each scholar establishes the value to be given to the seeds based on his supposition. In technical terms, we say that he hypothesizes a numerical system, then he replaces each seed with the value he has assigned to it (just as our decipher replaced symbols with digits) and reads numbers that certainly exist, but it is not at all certain that they are the numbers that Guamán Poma intended to represent on the yupana.

It is therefore understood that to decipher a mathematical message, it is not enough to read numbers. Instead, it is necessary to use a different parameter. What will our decryptor friend do?

Our deciphering friend, realizing that, whatever symbol-digit connection he makes, he would always be able to read "existing" numbers, would be forced to ask us the following question: "Are there only randomly chosen numbers in this encrypted message, or are they linked by logic?" Because if they are numbers chosen at random, it is useless for him to keep trying to decipher, because even if he found the right symbol-digit correspondence, he could not notice it, and any decipherment could be the right one. If, on the other hand, the numbers were linked by an algorithm, by a mathematical logic, then he could continue deciphering until he found a logical connection between the numbers read.

If we confirm that the numbers are linked by a calculation algorithm, then our friend will check if there are logical connections between the numbers read in each of the possible decryptions, and when he realizes that in one of the decryptions, the last number is the sum of all the previous ones, he will stop and say that he has found the solution. Again, given the large number of digits involved, he will not make any further attempts to decipher him, as a second solution is unlikely.

If, on the other hand, our friend had stopped at the first attempt at decipherment, saying that he had found the right decipherment because he could read numbers, he would have behaved in a manner equivalent to that of stopping at the first attempt in the linguistic field. In the linguistic field, the error is more evident because we "read" strange, non-existent words, while in the mathematical field we can be deceived by being able to read existing numbers, but we are still in error.

Moving on to the yupana, we can therefore say that it is not enough to read numbers on Guamán Poma's yupana to be able to say that we have deciphered it. It would be like stopping at the first attempt at deciphering seen in the linguistic context where only strange sounds were read. It would make no sense to do so, and trying to give meaning to the numbers read would not be convincing. To be able to say that we have succeeded in deciphering it, we have to read numbers linked together by mathematical logic on the yupana. For example, we must read an addition, a multiplication, etc., just as Locke (1912) read an addition on quipus.

More precisely, just as Locke was able to identify the right numbers to assign to each knot on the quipu to read an addition, in the same way, we must be able to assign the right values to the black and white circles on page 362 (360) of *Nueva Corónica* to read a mathematical message (Fig. 7.6). Presumably, it will be one of the four arithmetic operations.

Consequently, if only random numbers were inserted on Guamán Poma's yupana by the designer, decipherment would not be possible, because even if the numerical

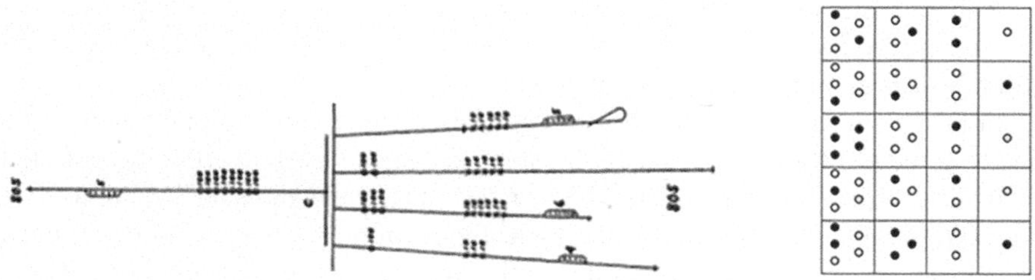

Fig. 7.6 Locke's quipu (left); image in the public domain and Guamán Poma's yupana (right). (Image by the author)

system used were centered by pure luck, the hypothesis could not be proved, just as the decipher would not have been able to tell what the correct symbol-digit correspondence was if the numbers in the encrypted message had been random. But if Guamán Poma represented a calculation, we might be able to read it, and it would finally shed light on Inca mathematics.

Let's now look back at the studies on Guamán Poma's yupana shown earlier, in light of the comments made.

All of the studies shown return either a single number on the yupana, or numbers that are not linked together by a calculation algorithm, so if Guamán Poma represented a calculation it follows that none of them were able to decipher them. If, on the other hand, we consider that Guamán Poma wrote only random numbers, the hypotheses of the scholars are all possible hypotheses, but not provable, so technically the number of errors present in their interpretations can vary between 0 and 55 out of the total of 55 balls, just as our hypothetical decipher friend by replacing the symbols with digits could not understand how many errors he had made: He couldn't notice it.

It is important to underline, however, that if a different study manages to identify an arithmetic operation on the yupana of Guamán Poma, all the previous hypotheses shown by the various scholars must be automatically discarded, as being able to read an operation means identifying a logical scheme that rigorously explains why the 55 symbols on the yupana have those features, are in those positions, and how they relate to each other. It is evident that given the large number of variables: 55 circles, of 2 colors, arranged in 20 squares, on 5 rows and 4 columns, in the repetitive sequence 1, 2, 3 and 5, outline a complex algorithm. The possibility that Guamán Poma wrote numbers at random and that by an incredible coincidence they can also be read as an ordered mathematical message representing an arithmetic operation, is not to be considered. It would be like thinking that if we take 55 toothpicks and drop them on a table, we will accidentally form a perfect geometric figure. Reading an arithmetic operation on Guamán Poma's yupana would therefore give the certainty of decipherment.

So I hope I have answered the first question clearly: why can everyone read numbers on yupana and what should be the correct numbers.

It should also be noted that some of the hypotheses listed on the interpretation of the yupana are not convincing a priori, as they follow mathematical logics very far from those present on the numerical quipu.

In fact, we have to consider that the calculations were carried out on the yupana and then the result was recorded on the quipu, so there were two different tools and methods to represent the numbers: seeds on the yupana and knots on the quipus. It is therefore reasonable to think that the logical transition from yupana to quipu must have been very simple: why build a complex passage that would tire the *quipucamayoc*? In this logic, the methods of De Pasquale (2001) and Kak (2014) would be discarded immediately as they provide for the base 40 and 144 respectively on the yupana, while on the quipu the trend is decimal, so after having carried out the calculations on the yupana, the *quipucamayoc* would have had to make another calculation to transform the base number 40 or 144 obtained on the yupana into another base 10 number in order to be able to record it on the quipu. In the same way, Wássen's (1931, 1940) hypothesis with the trend 1, 5, 15, 30 would be tiring in inserting numbers on the yupana for a mind accustomed to a decimal trend on the quipu. Chirinos Rivera's (2010) hypothesis, which predicts a different value for each location, in addition to the complex way of reading the numbers, is too uneconomical.

As for the second question—why everyone is able to perform correct calculations on the yupana but in a different way and which one is the right one—the answer will be shorter as well as being already guessed for what was said before: Once a scholar has taken his number system at face value, he can always skillfully construct a method of calculation by addition, subtraction, etc. All we need is mathematical skill to do this, just as in ancient times each people built different algorithms, but the results were obviously the same. 2 + 2 must always give 4, however the numbers are represented and however the logical path to make an addition is created.

So to be sure that we have identified an algorithm actually used on the yupana by the Incas, we have to be able to read it on the yupana designed by Guamán Poma on page 362 (360) of *Nueva Corónica*, and not build it from scratch, as we can build many different ones.

Ultimately, we can be sure that we have deciphered the yupana only if we can read on the yupana on page 362 (360) of *Nueva Corónica*, with that particular arrangement of black and white circles, a calculation algorithm, and if Guamán Poma had written only random numbers, deciphering it would be impossible.

Aware of this need, I would not have presented a work that had not identified a calculation in the yupana, I would not have shown an unprovable hypothesis of a sequence of random numbers that would have only lengthened the list. In the following, we will try to show that a multiplication is represented on the yupana, which is immediately visible as soon as the hypothesis of the use of an additive numerical system for powers of 10 is inserted. The only drawback is the presence of an error, a single incorrect symbol out of the 55 symbols used. This error will also be discussed later (in Chap. 8).

Competing Interests The author has no conflicts of interest to declare that are relevant to the content of this chapter.

References

Chirinos Rivera, A. (2010). *Quipus del Tahuantinsuyo: Curacas, Incas y su saber matemático en el siglo XVI*. Editorial Commentarios.

De Pasquale, N. (2001). Il volo del condor. *Rivista Dell'Ordine Degli Ingegneri Della Provincia Di Pescara*, 0, 1–19. http://www.quipus.it/DePasquale_Volo_del_Condor.pdf

Guamán Poma, F. de A. (1615). *El primer nueva corónica y buen gobierno*. Museum Tusculanum Press.

Kak, S. (2014). Guamán Poma's yupana and Inca astronomy. Preprint. In *arXiv*: Vol. 1401.7637. https://arxiv.org/ftp/arxiv/papers/1401/1401.7637.pdf

Locke, L. L. (1912). The ancient quipu, a Peruvian knot record. *American Anthropologist*, *14*(2), 325–332.

Wassén, H. (1931). The ancient Peruvian abacus. *Comparative Ethnographical Studies*, *9*, 189–205.

Wassén, H. (1940). El antiguo ábaco peruano según el manuscrito de Guamán Poma. *Ethnologiska Studier*, *11*, 1–30.

Chapter 8
Translating Guamán Poma's Yupana as a Multiplication

Abstract In this chapter, we build a logical path to deciphering Guamán Poma's yupana. First, we evaluate all the available material on the customs and traditions of the Incas. Then, we use a technical approach to produce an analysis that is as objective as possible. The result is convincing and self-demonstrative. The result also shows how to apply multidisciplinary data successfully. This observation is perhaps even more important than the decipherment itself for the future of archaeology and the history of mathematics and technology of ancient peoples.

Keywords Yupana · Decipherment

Deciphering the yupana of Guamán Poma (1615) means starting with the assumption that the drawing represents a calculation—that is, a coherent mathematical message—and not just random numbers. Otherwise, as was mentioned before, decipherment would be impossible because, regardless of the value assigned to the black-and-white counters, numbers can always be read on the yupana. However, unless we find a mathematical message, we cannot be sure that the numbers were the ones Guamán Poma intended.

Accordingly, we must ask ourselves the following question: is it too optimistic to think that Guamán Poma represented a particular calculation, or can we support the hypothesis on the basis of reasonable data?

What we can see is that the drawings of *Nueva Corónica* are always very detailed, and this has particular importance for our specific analysis. For example, we note that when Guamán Poma represents a scribe as writing sentences on a sheet of paper, we can read the lines of writing. There are five cases of this in the *Nueva Corónica* (Figs. 8.1, 8.2, 8.3, 8.4 and 8.5):

We observe that when actual messages are drawn, even if on very small sheets of paper that are upside-down from the perspective of the readers, Guamán Poma reports a coherent and readable message. This gives us hope that the mathematical message on page 362 (360) will also be coherent, especially when we consider that the yupana is much larger than the sheets of paper being written on by the various scribes; the yupana is also oriented towards the reader, deliberately in sight (Fig. 8.6).

C. Florio, *The Mathematical Enigma of Guamán Poma's Yupana*,
https://doi.org/10.1007/978-3-031-92662-4_8

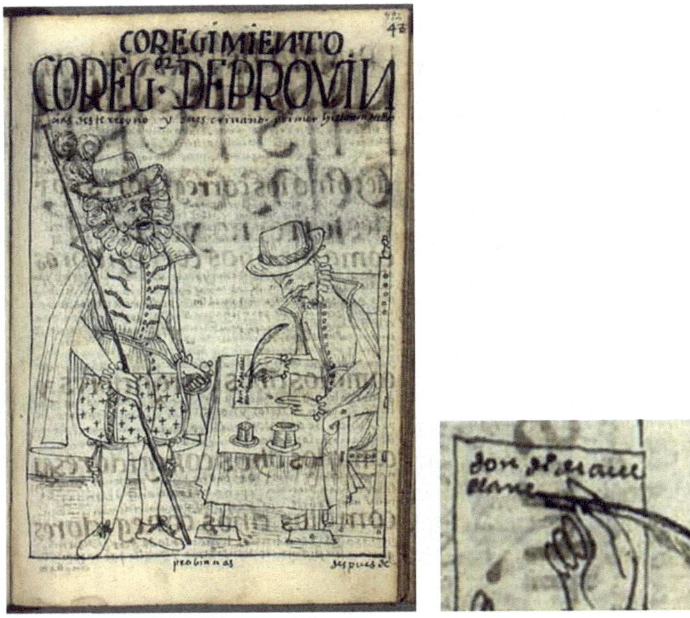

Fig. 8.1 Page 492 (488) of Guamán Poma's *Nueva Corónica* with detail enlarged to show the legible writing. The line reads, "don Diego de Auedano." (Image in the public domain)

Fig. 8.2 Page 525 (521) of Guamán Poma's *Nueva Corónica* with enlarged detail. The line reads, "don Diego de Mendoza, corregidor desta prouin." (Image in the public domain)

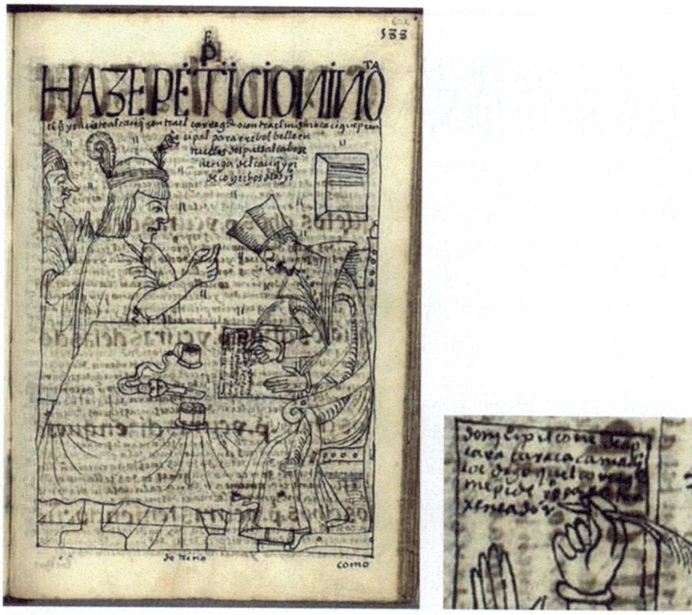

Fig. 8.3 Page 602 (588) of Guamán Poma's *Nueva Corónica* with enlarged detail. The line reads, "Don Juan Pilcone de Apcara, curaca camachicoc: Digo que el corregidor me pide yndio para traxeneador." (Image in the public domain)

Fig. 8.4 Page 784 (770) of Guamán Poma's *Nueva Corónica* with enlarged detail. The line reads, "Del terzio de San Juan pagò 80 pesos Pedro N." (Image in the public domain)

Fig. 8.5 Page 828 (814) of Guamán Poma's *Nueva Corónica* with enlarged detail. The line reads, "En el nombre de la Santícima Trinidad hago el testamento de don Pedro." (Image in the public domain)

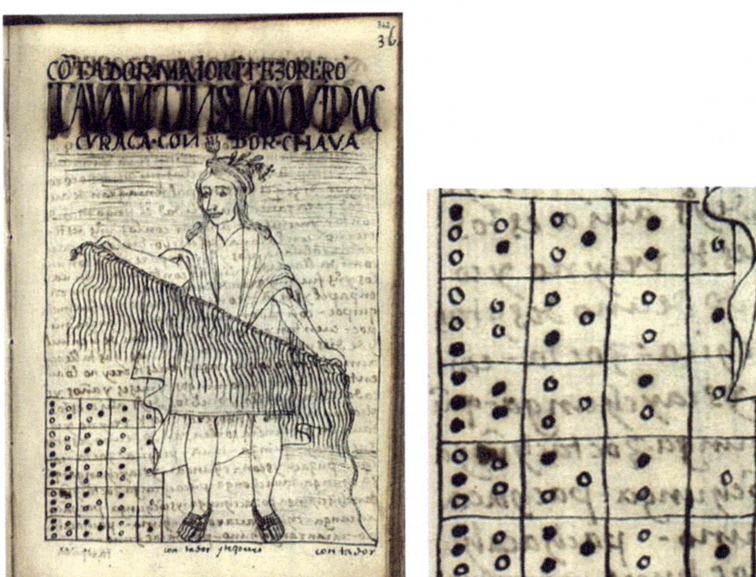

Fig. 8.6 Page 362 (360) of Guamán Poma's *Nueva Corónica* with enlarged detail. (Image in the public domain)

These circumstances suggest that, just like the legible lines of writing, the yupana will likewise bear a legible message.

Once we accept the idea that the yupana represents a calculation, we must next ask what kind of calculation it could be. The *quipucamayocs* in the Inca empire kept very detailed accounts, and they reported everything to their superiors in a pyramid hierarchy that reached the Inca. They performed an annual census of the population, took into account the stocks of food, the working days of each individual, and so on. They were careful and scrupulous accountants, rather than mathematical scholars. In addition, according to what the Chroniclers reported, the calculation represented on the yupana must reasonably be one of the four arithmetic operations (addition, subtraction, multiplication, and division).

Beyond simple arithmetic, the most we might hypothesize would be some sort of percentage calculation. Given only simple arithmetic, the question immediately posed is this: How it is possible that today—an era in which mathematical logic has reached sublime heights—we have not yet identified the simple arithmetical message of Guamán Poma's yupana?

The only possible answer to this question is that our analysis is "clouded" by our mathematical training. Our mindset—our way of conceiving numbers and algorithms—is undoubtedly different from that of the Incas. This means that we look at the yupana from the wrong angle, and that every kind of analysis we do is conditioned by factors and preconceptions of which we are likely not even aware. Our mathematical training, for better or for worse, strongly shapes our minds.

To study the yupana, the present author therefore asked this basic question: If it is already difficult to discard the preconceptions that we know well—the ones that, for example, we acquire during our childhood and reject in our adulthood through the powers of adult reasoning—how can we possibly discard the preconceptions that prevent us from reading the logical message of the yupana, when we don't even know what they are? Preconceptions are a paralyzing handicap in research. All of this led to the insight that it was useless to continue to look at the yupana in order to decipher it; only a brilliant and free mind, one capable of going beyond its conditioning, could possibly succeed. Another kind of approach was required.

In science, particularly in the field of physics, whenever an object cannot directly be observed, what is happening around it is analyzed in order to gain information about the unobservable object. In trying to follow this path, the present author imagined a *quipucamayoc* performing a calculation on a yupana. The calculation to be carried out was, for example, a simple multiplication: $15 \times 18 = 270$. However, the yupana itself was mentally obscured in order to focus on everything that would happen around it for the purposes of calculating.

From the eyewitness accounts given by various colonial chroniclers, we can imagine that the *quipucamayoc* would have taken kernels of corn and arranged them in some way on the yupana to represent the initial data he would need for his calculation; he would then move them, with great skill, in carrying out the calculation. We do not yet try to understand how he arranges or moves the kernels, but we observe that at the end of the calculation, he stops and reads the result, 270, which is represented on the yupana with kernels in a particular arrangement, not yet known to us.

At this point, he records the number 270 on a quipu, making two knots in the area for hundreds, seven knots in the area for tens. We can, therefore deduce that the *quipucamayoc* could represent the number 270 (and obviously, all other numbers) in two different ways: He could use kernels of corn on the yupana and knots on the quipu. Accordingly, we find ourselves confronting two different ways to represent numbers used by the same people, which suggests that the transition from one to the other would necessarily have been simple and intuitive.

When we look for a connection between kernels and knots, the most immediate and intuitive connection is one of direct correspondence between them, so a kernel would correspond to a knot. But then we encounter the problem that the knots do not have the same value. In fact, depending on the height on the rope at which they are knotted, they correspond to the values 1, 10, 100, and so on. This suggests that the kernels should also have a variable that associates them with the different powers of 10. At this point, the two colors of the counters in Guamán Poma's drawing comes into play. They suggest that the difference in kernel color might indicate the different powers of 10. In support of the hypothesis, we recall the words of Juan de Matierzo, who wrote in 1567 that the Incas performed their calculations with *colored* stones.

Color, therefore, informs an alternative hypothesis, and this inspires us to continue in this direction.

The hypothesis of color is fascinating, but it immediately poses two problems that at first glance are very substantial:

1. If the black-and-white counters represented two particular powers of 10 (for example, 10 and 100), how were the other powers of 10 represented? Certainly, two colors would be insufficient for representing all of the powers of 10 named and used by the Incas.
2. For the drawing, researchers talk in terms of numerical counters, but in the Chronicles, various observers wrote that kernels of corn were used on the yupana. Corn is yellow, so with only a single color, the hypothesis would collapse immediately.

In reality, these two problems are false and are quickly dispelled. The first question can be answered by simply looking at the *Nueva Corónica* (Fig. 8.7):

It is a black-and-white text—black ink used on white paper—and this means that if Guamán Poma had wanted to represent numerical counters in various colors, he could only have drawn them in black or white. Yet the possibility has never been considered—even if, once pointed out, it seems trivial—that if the text is in black and white, the counters in reality could also have been colored, and they might even have included more than two colors.

The second question is more difficult to answer because in Europe, we think only of the color yellow whenever we talk about corn. However, here too the answer is simple: In South America, corn comes in a variety of colors (Fig. 8.8), including white, yellow, green, purple, red, blue, and black (Salvador-Reyes and Silva Clerici 2020; Staller 2010).

Fig. 8.7 The *Nueva Corónica* is a black-and-white text. (Image by Palnatoke and distributed under a creative commons license)

Fig. 8.8 South American corn, showing a variety of colors. (Image by Logga Wiggler via Pixabay and distributed under a royalty-free license)

This shows that a people cannot be studied in impermeable compartments: Its history, culture, technology, art, language, and territory are all closely linked.

Having accepted the hypothesis that the counters on the yupana, depending on their color, can be worth 1, 10, 100, and so on, we will insert these data into the drawing and then see if we can read an arithmetic operation.

For simplicity and in order to have small numbers to manage, we will assume that the black counter is worth 1 and the white one is worth 10 (Fig. 8.9). Obviously

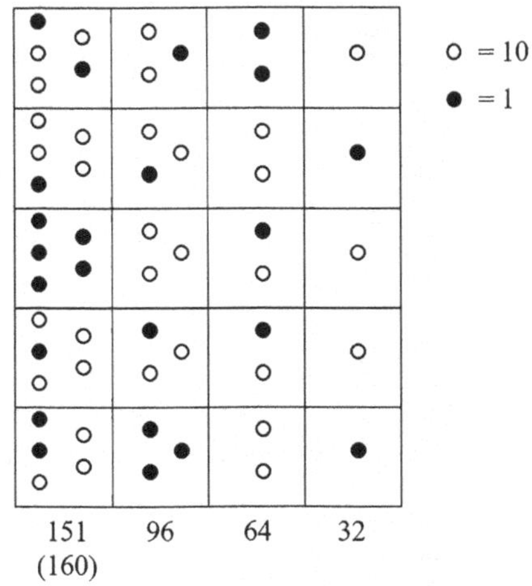

Fig. 8.9 Guamán Poma's yupana, with black counters assigned a value of 1, white a value of 10. (Redrawn by the author from an image in the public domain)

different choices can be made, but the logic of reading the yupana will remain the same:

We will start with the rightmost column with one counter in each square. If we add the values of three white counters worth 10 each and two black counters worth 1 each, we obtain the number 32.

Now, we will move one column to the left, to the column with two counters per square. Here there are six white and four black counters. Using the valuation assigned, we obtain the number 64. With the results of two columns, we already perceive a specific numerical order: a number (32) and its double (64).

We will move another column to the left, to the column with three counters per square. Here there are nine white and six black counters. This gives us the number 96, and again confirms a specific numerical order: we now have a number (32), its double (64), and its triple (96).

All this leads us to expect that in the leftmost column with five counters, we should have five times as much as 32, that is, $32 \times 5 = 160$. However, when we count the counters in the leftmost column, we are disappointed because there are 14 white counters and 11 black counters. On the valuation assigned, their sum yields: $(14 \times 10) + (11 \times 1) = 140 + 11 = 151$.

However, when we examine the leftmost column closely, we realize that the result would be 160 if one black counter were instead white. This suggests a possible error. The possibility of an error is reasonable when we take into account the fact that calculation errors are not uncommon in the history of mathematics.

To be precise, this cannot be considered an error in calculation, but rather, an error in drawing. After all, we are not analyzing an Egyptian papyrus, a Babylonian clay tablet, or an Incan quipu, but rather, a drawing of a calculation. While the yupana itself was an object on which kernels of corn were placed and then moved to calculate, what Guamán Poma has provided to us is a drawing of a yupana with kernels.

We should also note that the mistake was made in the column with the largest number of counters. This means it would have been the most difficult to check visually. The figure of the *quipucamayoc* holding a quipu also appears to have been drawn first, with the yupana added to the background. Because the larger drawing had already been completed and a mistake could not simply be erased, correcting it would have meant redrawing it from scratch. This would have been time-consuming, since the figure of the Contador would also have to be redrawn.

It is possible to arrive at the number 160 on the last column through a probabilistic analysis. In detail, the analysis would proceed as follows: In the first column on the right, there are five counters in total, so at most six numbers could have been represented: all black counters, all white, one white and four black, two white and three black, three white and two black, or four black and one white. Given six numbers in the first column, it is possible to build only six triples of numbers in the first three columns that meet the condition of indicating a number, its double, and its triple. However, it is possible to write up to 11 different numbers in the column with two counters per square, and 16 in the column with three counters per square. This means that all the possible triples that can be identified for the first three columns starting from the right are 6 × 11 × 16 = 1,056.

The probability, then, that randomly drawing black-and-white counters in the first three columns would produce a number, its double, and its triple is 6/1056, or about 6/1000. For a more complete analysis, it would be necessary to evaluate all possible arrangements for each group of counters that fill the column for each number considered. The calculation becomes more complex and is reported in the appendix; the result is about 20/1000, or 2/100.

A simpler calculation can also be made. As the starting point of the analysis, consider that the rightmost column contains the number 32 (on the values assigned). Then ask, what is the probability that the next two adjacent columns contain its double and triple? The probability is 3/100 (as reported in the appendix), so it is fully comparable with the previous one.

All this means that if the counters were inserted at random on the first three columns from the right, they would be highly unlikely to produce a number, its double, and its triple, having in this case a very low probability of being realized (2/100 or 3/100) compared to the much more likely case of being disordered (98/100 or 97/100). This suggests that the counters were not drawn randomly, but with the precise desire to indicate a number, its double, and its triple. This supports the likelihood that the column of five counters contains an error, with respect to the logic of the sequence.

If we consider the hypothesis of error reasonable in the light of its probability analysis, we can recognize in Guamán Poma's yupana a multiplication. When we translate it into our mathematical language, we would write it like this:

$$32 \times 5 = 32 \times (2+3) = (32 \times 2) + (32 \times 3) = 64 + 96 = 160.$$

Let us see how this multiplication is displayed on the yupana (Fig. 8.10):

1. The rightmost column, where there is only one counter per square, contains the multiplicand, in this case, the number 32. The five necessary counters (two black and three white) are then inserted, one per square. Therefore, five rows are used.
2. The multiplier is the number 5, and as five kernels of corn, this number can be subdivided into smaller parts; in this case, Guamán Poma chose 2 and 3. $(2 + 3) = 5$.
3. The numbers chosen (i.e., 2 and 3) represent the number of counters that will be inserted into the multiplier columns adjacent to the multiplicand (rightmost) column.
4. Into the column of 2 are inserted the counters that double the first column: 4 black counters and 6 white counters. In effect, a partial multiplication takes place:

$$32 \times 2 = 64.$$

5. Into the column of 3 are inserted the counters that triple the first column: 6 black counters and 9 white counters, carrying out the partial multiplication $(32 \times 3) = 96$. However, the counters in each square are not aligned in a row, but instead appear as 2 on the left and 1 on the right. Since the columns of 1 and 2 would have already been filled in on the yupana, this suggests that the column of 3 shows the sum of columns 1 and 2 (that is, $(32 + 64) = 96$), rather than the partial multiplication (that is, $(32 \times 3) = 96$). The juxtaposition of the two columns, one next to the other, seems likely to symbolize the sum of the two adjacent columns.
6. In the last column, a sum takes place $(64 + 96) = 160$; that is, the counters of the columns of 3 and 2 are added and the product is obtained. In this case, the counters again appear as 3 on one side and 2 on the other, confirming that that arrangement found in the column of 3 really does represent a sum.

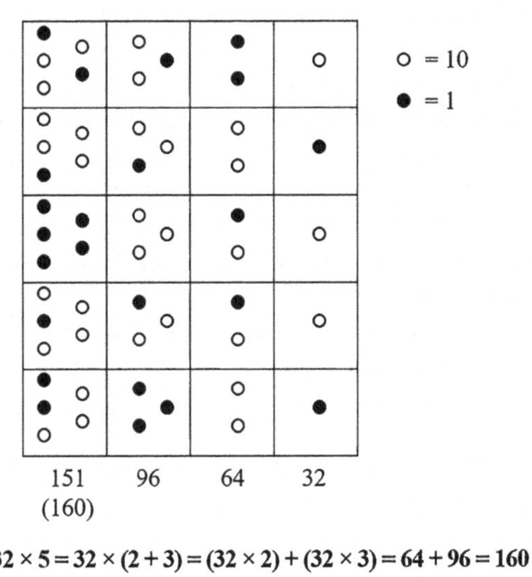

Fig. 8.10 Florio's interpretation. Guamán Poma's yupana shows the multiplication of 32 by 5. (Image by the author)

Other examples of multiplication using this calculation algorithm are shown in Chap. 11.

Identifying a multiplication in the yupana, even if it contains an error, confirms the proposed decipherment, and further, supports the hypothesis of an error in the final column.

Here is the explanation: on the yupana, there are two interlocking logical sequences: the first is the sequence of a number, its double, its triple, and its quintuple (with an error). The second is the multiplication that identifies the multiplicand, the multiplier divided into two terms, the partial multiplications of the multiplicand by the terms in which the multiplier is divided, and then the final sum; the result is formed by 25 symbols (the counters in the column on the left) with an error (only one wrong counter out of 25). So the multiplication carried out using 55 counters, overall, has only one error out of 55.

These two structures are interlocking, in the sense that a multiplication would not necessarily have been read if the initial sequence, however orderly, had been different.

Let's imagine that instead of the sequence 1, 2, 3, 5, we had the sequence 1, 2, 3, 10. Even if in each column we had the right black and white counters to indicate, for example, the number 32 taken one time, two times, three times and ten times (instead of five as in Guamán Poma's drawing), this sequence would not be read a multiplication. However, neither would it be read as another operation; instead, it would only be an ordered sequence.

The identification of the multiplication in the sequence 1, 2, 3, 5 is, therefore, a double confirmation that we have correctly deciphered the numbers in the four columns—that is, that we have assigned the right values to the black and white counters. The complexity of these two interlocking structures, in creating a coherent mathematical message, implies logical thinking and not randomness in the arrangement of the counters. Thus, it is very reasonable to support the error hypothesis. In the same way as was shown in the sentence in Fig. 7.3, it seems plausible that the decipherer would take the decipherment at face value and consider the error to be a normal typo.

The presence of a logical structure that correctly takes numerous parameters into account and amply justifies the presence of an error thus leads to the reasonable conclusion that the mathematical message has been correctly deciphered.

On the other hand, even if this decipherment were rejected, it nonetheless suggests that in the future, the only decipherments accepted as valid will be ones reporting an algorithm, whether they are exact or contain errors.

It no longer makes sense to suggest, as previous scholars have, interpretations of the yupana that yield random numbers. Occam's razor would inexorably cut these hypotheses in light of finding a multiplication in Guamán Poma's yupana, even one that suggests a mistake in drawing.

It is conceivable that another valid decipherment might someday be found, perhaps even one whose algorithm contains no errors. However, this seems highly unlikely. Why this is so can be illustrated with an example: suppose we have the sequence of numbers (12, 2, 14). If we interpret this sequence as a subtraction

(12 − 2 = 10) with an error in recording the second digit of the number 14, it could be argued instead that the algorithm was an addition without an error (12 + 2 = 14) or a multiplication with an error (12 × 2 = 24) on the first digit of the number 14, which should have been 2 and not 1. This example involves only three numbers, and five digits altogether. Consider how much more difficult it would be to produce a similar example for Guamán Poma's yupana, given that it contains 55 symbolic numbers in a particular distribution (single, double, triple, and quintuple). Anyone who can find another such algorithm in the yupana, even one with two or three errors, must sincerely be congratulated.

I hope that in the future, we might find another drawing of a yupana similar to the one drawn by Guamán Poma, as this would further confirm this decipherment. Perhaps in a library or repository somewhere, there are still some lost notes or travel diaries written by Jesuit missionaries who worked in the Incan Empire after the Conquest, waiting to be rediscovered and published. Time and future research will tell.

Let us now see how this decipherment sheds a different light on the shape of the yupana and the numerical system used by the Incas considered so far.

Competing Interests The author has no conflicts of interest to declare that are relevant to the content of this chapter.

References

De Matienzo, D. J. (1567). *Gobierno del Perú. Obra escrita en el siglo XVI por el licenciado Don Juan Matienzo*. Compañia Sud-Americana de Billetes de Banco.

Guamán Poma, F. de A. (1615). *El primer nueva corónica y buen gobierno*. Museum Tusculanum Press.

Salvador-Reyes, R., & Silva Clerici, M. T. P. (2020). Peruvian Andean maize: General characteristics, nutritional properties, bioactive compounds, and culinary uses. *Food Research International, 13*, 108934.

Staller, J. (2010). *Maize cobs and cultures: History of* Zea mays L. Springer Science & Business Media.

Chapter 9
The Implications of Detecting a Multiplication in Guamán Poma's Yupana

Abstract Deciphering the yupana allows us to trace its shape and size, determine the numerical system used by the Incas, and more accurately interpret the numerical quipus.

Keywords Yupana · Number systems

1. Identifying a multiplication means that Guamán Poma's (1615) drawing reported an actual calculation, and this in turn implies that the checkerboard yupana is the real counting board used by the Incas. This means that the three-dimensional object called the archaeological yupana was not the object used by the Incas to carry out calculations. It can no longer be considered a yupana.
2. It is clear from its decipherment that the number of rows used in multiplication was equal to the number of counters needed to indicate the multiplicand in the first (rightmost) column, while the number of interim columns depended on the multiplier selected and the parts into which it was divided, to which were then added the initial and final columns. So the yupana as recorded by Guamán Poma seems to report only one portion of the yupana, the one with the squares filled with the kernels needed for that particular multiplication. In reality, the yupana would have had more rows and columns in order to support calculations with different numbers. We can also speculate that, in order to avoid carrying a very large yupana, the *quipucamayoc* might have used many small yupanas, combining them as needed.

   Consider that for a multiplicand like the number 98, a yupana with 17 rows would be needed (nine counters for the tens and eight for the units, all comprising the first column). This seems to be precisely confirmed in two pages of the *Nueva Corónica*. The first, as has already been mentioned, is page 363 (361), where Guamán Poma wrote that the Incas used tables—"cuentan en tablas," plural—to make their calculations. The second is the page 804 (790), where the kneeling man appears to support several small, overlapping checkerboard yupanas with his left hand (Fig. 9.1). This might suggest that the yupana on page 362 (360) was represented on an enlarged scale, the better to show the calculation.

C. Florio, *The Mathematical Enigma of Guamán Poma's Yupana*,
https://doi.org/10.1007/978-3-031-92662-4_9

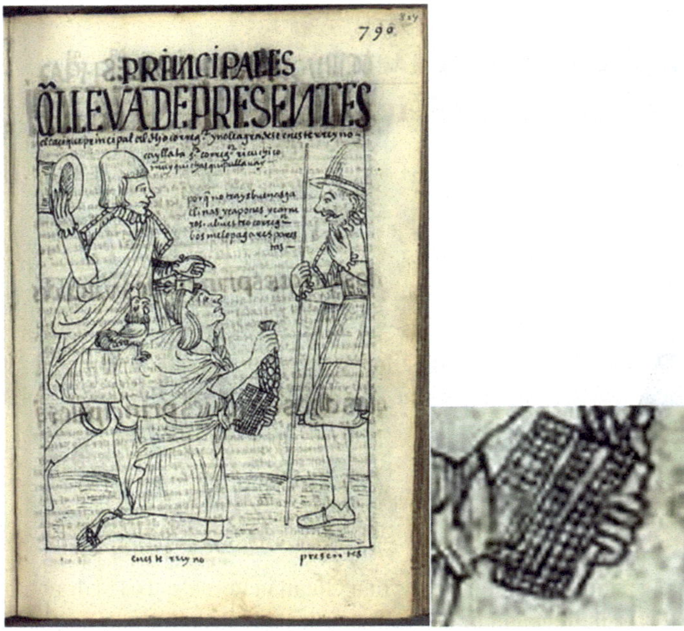

Fig. 9.1 (Left) Page 804 (790) from the *Nueva Corónica*. (Right) Detail from the same page, showing what could be multiple small yupana. (Images in the public domain)

Unfortunately, Guamán Poma made no reference to this particular image page 804 (790) in his text, so we are left without an explanation.

Alternatively, the object held by the kneeling man might be a folded bolt of cloth, fabric, or netting. However, the edges are quite square, and in other drawings by Guamán Poma, the fabrics on clothes or tables invariably have a softer appearance, with wrinkles and rounded edges. In addition, each of the boards held by the man has a white border, whereas if it had been a fabric or netting, it would likely have been completely checkered. Scholars of the Andean world might find it interesting to explore the issue further.

3. Now that we have identified a multiplication algorithm, we can see that the sequence of one, two, three, and five counters in four columns was not fixed, but rather, depended on the particular multiplication being carried out. In particular, the first (rightmost) column always contained one counter per square, as it represented the multiplicand. The multiplier could then be divided into smaller parts as needed, generating more columns whose number of counters per square depended on both the value of the multiplier and how it was divided. Finally, in the last (leftmost) column, the number of counters per square was the sum of the number of counters per square of the multiplier columns. Thus, the multiplication shown in Guamán Poma's drawing did not represent the Fibonacci sequence; instead, it only coincidently uses the numbers 1, 2, 3, and 5. That is, in the column with one counter per square, we insert 32. The multiplier is 5 and is divided

Fig. 9.2 Hypothetical yupana without holes. (Diagram by the architect Eliana Continanza and used with permission)

into 2 + 3, yielding columns with two and three counters per square. Finally, in the leftmost column where the final sum is displayed, the multiplier columns are added, yielding squares with five counters each.

This confirms that there were no holes on the yupana in sequence 1, 2, 3, 5 or in any other sequence: the surface of the yupana was smooth. Thus, the yupana was a smooth board, without any holes, with horizontal and vertical lines in the manner of a chessboard (Fig. 9.2).

Into each square was inserted the appropriate number of kernels of particular colors. These boards were likely made of wood so that they would have weighed little, making them easily transportable. The reason why there are no archaeological finds of this object may be due both to the deterioration of the material and to the possibility it was used as fuel. However, the most reasonable hypothesis is that not understanding their nature, they may have been mistaken for checkered wooden boards that could be pieces of furniture. Consider that the checked pattern was also a decoration widely used on Inca fabrics. It is therefore possible that there are still artifacts preserved in museum deposits, but still not identified as yupana.

On the other hand, they were looking for a yupana with five rows, four columns, and 55 holes, which in reality does not exist.

By identifying multiplication on the yupana, and then clarifying its appearance, we have answered the questions posed in Chap. 4.

Let us now also return to the question asked in Chap. 3: Is the quipu system additive or positional? Because we were able to read a multiplication on the yupana, it means that we gave the right values to the seeds, in the same way that Locke gave the right values to the knots on the quipu by being able to read an addition.

To give a visual image that is perhaps more intuitive, we can compare the deciphering of the quipu and yupana to the opening of the door of a safe after entering the right code, or to the access to a PC after entering the right password: Locke's addition "jumped out" at us. It was visible only after he had assigned the correct values to the knots. In the same way, on the yupana, multiplication became visible only after assigning the right values to the seeds.

In technical terms, to say that we have assigned the right values to the seeds and knots, means to say we have identified the numerical systems with which the numbers are represented on the yupana and quipu, respectively. In particular, the decipherment of multiplication on yupana is obtained by inserting a numerical system that associates the seeds with the values 1, 10, 100, etc., depending on their color.

As was explained in Chap. 2, this type of number system is additive for powers of 10, in fact the seeds have a value determined by a precise characteristic (color) and not by their position. For example, if the white seed is worth 10, it will always be worth 10, regardless of the square of the yupana in which it is inserted. So the yupana can't be a positional system, with or without the knowledge of zero (that's why there are no empty squares within the calculation scheme on the yupana, which happens in the Chinese abacus that uses a positional system without the knowledge of zero).

The values of the seeds contained in a column are only added together, just like the Egyptian symbols, and they can be moved along the column from one square to another, but the number that will be read will always remain the same. Finally, this system does not use zero: the need to associate a colored seed with zero did not arise because, in an additive system, it makes no sense to add nothing.

This becomes more clear if, for the moment, the horizontal lines are deleted from the yupana (Fig. 9.3).

In this hypothetical version of the yupana, the counters inside each column can be moved as needed, but the number represented remains unchanged, as was true when the Egyptian symbols were reordered. Thus, the lines of the yupana do not serve to indicate positional value; instead, they enable the counters to be controlled visually when filling in the squares, thereby minimizing the potential for insertion errors. On Guamán Poma's yupana, therefore, there is an additive decimal system, in which zero is not used (that is, there is no seed of a particular color for zero), and the calculations are voluminous and without empty spaces as for example.

We can look at Fig. 11.2, in which the multiplication 2,212 × 27 = 59,724 takes place with the Inca method. Because the Inca method uses an additive system, it requires 133 colored seeds. In comparison, the same multiplication in the Western positional system (Fig. 9.4) requires 24 digits (a ratio of about 6 to 1).

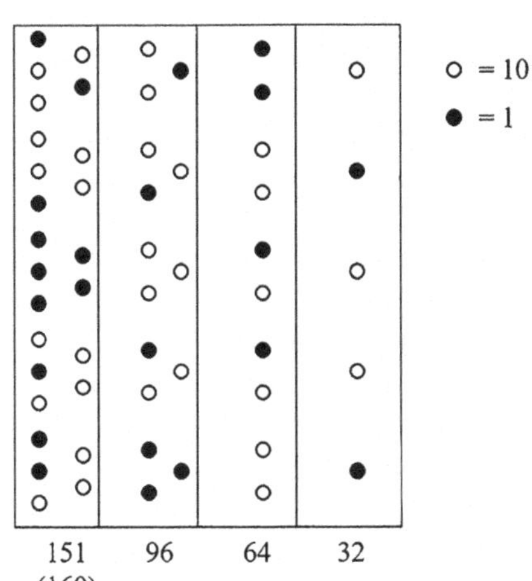

Fig. 9.3 Hypothetical version of Guamán Poma's yupana, with the horizontal lines removed. (Adapted from an image in the public domain)

Fig. 9.4 Multiplication 2,212 × 27 = 59,724. With the Western method, 24 digits are required. (Image by the author)

```
  1  1
   2212
 x   27
     1
  15484
 +44240
  59724
```

The identification of an additive system that requires voluminous calculations takes us back to a yupana with many squares and the possibility of using more than one square side.

The decipherment of yupana then sheds light on the number system used by the Incas: an additive system for powers of 10.

So we can finally answer the question posed in Chap. 3, where we did not know whether an additive or positional decimal number system was used on quipus, as the same numbers were read with both systems. The chapter concluded by saying that only by discovering the methods of calculation of the Incas on the yupana could one answer with certainty be given.

We can now say that the Incas used an additive system for powers of 10, and that consequently, each knot on the quipu has a precise value depending on its height and should not be multiplied by any coefficient; the number is obtained by summing the values of each knot; moreover, empty spaces do not indicate zero, but only the absence of knots of the value corresponding to that particular height. Zero did not exist for the Incas.

Finally, as anticipated, I would like to return to the work of Radicati di Primeglio (1979), to point out how very close he came to deciphering the yupana, without actually hitting his target, perhaps because he was not very expert in ancient numbering systems. If we look at his multiplication hypothesis shown in Fig. 6.14, we can see that he used a yupana with more than five rows and four columns, and when he inserted the seeds and there are too many to fit in a box, he made a substitution of ten black seeds with a single white seed. He writes precisely:

> El principal inconveniente que debió presentarse fue el excesivo amontonamiento de fichas en algunos casilleros. Sin embargo, estamos seguros de que este obstáculo fue superado fácilmente mediante el empleo de fichas de color distinto a las corrientes, para señalar conjuntos de estas últimas; no sería de extrañar, por ejemplo, que se indicase con valor de diez frijoles o maíces negros a uno blanco o viceversa, tal comolo hicieron los mayas al conceder valor de cinco marcas (frijoles o maíces) a una barra o palito de madera. pag. (Radicati di Primeglio 1979, pp. 295–296)
>
> The main inconvenience that had to occur was the excessive piling up of seeds in some boxes. However, we are sure that this obstacle was easily overcome by the use of tiles of a different color than the ordinary ones, to indicate sets of the latter; It would not be surprising, for example, if the value of ten black beans or corn were indicated to a white one or vice versa, just as the Mayans did when they gave a value of five marks (beans or corn) to a wooden bar or stick. (as translated by the author)

He was then able to hypothesize a yupana larger than the one drawn by Guamán Poma, as well as the use of seeds of different colors to represent the various powers of 10. But he didn't realize that he had the solution in his hands: on the yupana, the system was additive for powers of 10. On the other hand, as already mentioned (Chap. 4), Locke (1912) had also drawn, next to each knot on the quipu, its value, and he had not written the number zero next to the knotless zones. He, too, had touched on the hypothesis of the additive system on quipus. All this shows how education in the positional system has conditioned our minds, and how preconceptions are able to draw veils and obscure even very obvious things. All the other scholars were also conditioned and led "astray" by the further hypothesis of the 55 holes introduced by Wássen (1931, 1940).

Ultimately, we can conclude by saying that the Incas used an additive system for powers of 10. The calculations were carried out on the yupana, a smooth and square board, on which seeds of various colored species or colored stones were inserted: each color corresponded to a power of 10. The result was recorded on the quipu with knots at different heights, each height representing a power of 10. Each knot therefore had a precise power value of 10, depending on the height at which it was located. Adding up the values of all the knots read the number. Zero didn't exist.

Competing Interests The author has no conflicts of interest to declare that are relevant to the content of this chapter.

References

Guamán Poma, F. de A. (1615). *El primer nueva corónica y buen gobierno*. Museum Tusculanum Press.

Locke, L. L. (1912). The ancient quipu, a Peruvian knot record. *American Anthropologist, 14*(2), 325–332.

Radicati di Primeglio, C. (1979). *El sistema contable de los Incas: Yupana y quipu*. Libreria Studium.

Wassén, H. (1931). The ancient Peruvian abacus. *Comparative Ethnographical Studies, 9*, 189–205.

Wassén, H. (1940). El antiguo ábaco peruano según el manuscrito de Guamán Poma. *Ethnologiska Studier, 11*, 1–30.

Chapter 10
Preconceptions Preventing the Yupana from Being Identified as a Multiplication

Abstract This chapter lists the preconceptions born of our modern mathematical training that prevented us from reading a simple multiplication in Guamán Poma's yupana.

Keywords Yupana · Decipherment

This interpretation suggests that various preconceptions have complicated previous attempts to decipher the yupana. Now, however, we are able look at the yupana from an entirely new angle, and identify and then discard the now-outdated preconceptions:

1. The number of rows is not set to five, nor is the number of columns limited to four.
2. The counters can be of many different colors, not just black (with unfilled holes) and not only black and white.
3. The yupana contains no holes, but is instead a flat surface with horizontal and vertical lines in the manner of a chessboard.
4. Apart from the multiplicand (rightmost) column, which necessarily contains only one counter per square, the other columns do not contain a specified number of counters per square. Rather than the 2, 3, and 5 shown in Guamán Poma's drawing, there can be any number of counters per square, and there can also be extra columns with other numbers of counters per square.
5. The biggest preconception of all was the assumption that the yupana represented a positional decimal system. This choice was both reasonable and consistent with the interpretation of the quipu as a positional decimal system: Each knot (knot cluster) was multiplied by a particular power of 10 according to its position on the string, and an empty space represented a zero. The yupana was assumed to represent this same organization. However, its decipherment shows that the yupana instead represents an additive numerical system, one in which the counters are valued as the different powers of 10 according to their color, rather than their position (higher or lower) within a column. In encoding value with color,

C. Florio, *The Mathematical Enigma of Guamán Poma's Yupana*,
https://doi.org/10.1007/978-3-031-92662-4_10

the yupana counters are analogous to the Mesopotamian tokens (which encoded numerical value with size and shape) and Egyptian hieroglyph numerals (which encoded numerical value with shape). This, in turn, implies that the quipu has been similarly misinterpreted as positional, when it is more likely an additive system, albeit one that encodes value with position on the string (this was because the rope was chosen as the means to represent the numbers).

6. At the end of the calculation, the identified multiplication algorithm always yields a rectangular checkered structure filled with counters, built with side-by-side columns of the same height. This rectangle, with its regular, four-cornered shape that to us resembles a spreadsheet or an abacus, gave rise to the idea that the yupana on page 362 (360) of the *Nueva Corónica* was a fixed abacus with five rows and four columnstag. The idea that it was the pattern of a particular mathematical operation was not easily visible to us, as this does not happen in our algorithms, or it only happens in special cases. In general, the diagram does not have the shape of a straight quadrilateral, such as the one shown in Fig. 10.1:
7. As a final consideration, with respect to our way of thinking about numbers, we would be unlikely to distribute the yupana counters randomly, as Guamán Poma (1615) did (Fig. 10.2A); instead, we would likely arrange them in an orderly fashion like an Egyptian scribe would (Fig. 10.2B). An orderly arrangement would also be less likely to produce the mistake in drawing, as a black counter in a white line would have been quite noticeable (Fig. 10.2C).

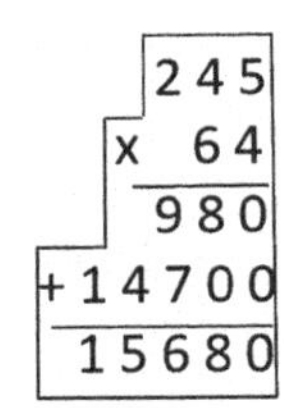

Fig. 10.1 The form that contains the Western multiplication scheme is neither a square nor a rectangle, as is the case in the Inca one. (Image by the author)

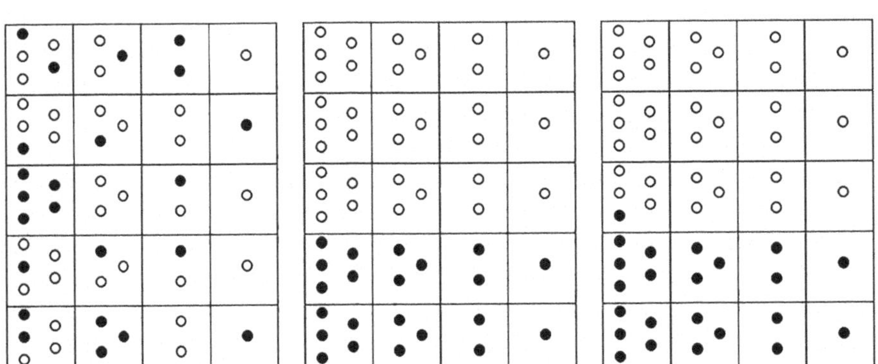

A, The original yupana with counters randomly ordered.

B, The yupana with counters reordered and the drawing mistake corrected.

C, Yupana with counters reordered and the drawing mistake uncorrected.

Fig. 10.2 Rearranging the counters on Guamán Poma's yupana. (Image by the author)

Why would Guamán Poma have used the disorderly arrangement that has made deciphering the algorithm it contained so difficult? Perhaps for this very reason: to make it difficult for the uninitiated to understand the algorithm; the knowledge of calculation methods might have been intended for only a select few. Or, more simply, it might have reflected a practical reason. Consider, for example, what would be involved in compiling the column of 2. In this column, the *quipucamayoc* had to insert 10 counters (six white and four black). If he had taken one counter at a time from the basket and placed it in the corresponding square, he would have had to turn his gaze first to the basket with the kernels, and then to the yupana, and then back to the basket, and so on, a bit like watching a tennis match. Doing this for all the counters needed would plausibly be boring and tiring. It is reasonable to think, therefore, that he might have collected all the counters to be inserted into a column at once, then arranging them randomly along the column from their being mixed in his palm. Essentially, any arrangement would have worked just well, since the same number would result in any case in an additive system. In drawing the yupana, Guamán Poma might have been inspired by the "messy" distribution he had often seen. Or the reason may be another unknown one; surely this messy arrangement did not help decipherment for many years.

Competing Interests The author has no conflicts of interest to declare that are relevant to the content of this chapter.

Reference

Guamán Poma, F. de A. (1615). *El primer nueva corónica y buen gobierno*. Museum Tusculanum Press.

Chapter 11
Potential Arithmetical Operations on Guamán Poma's Yupana

Abstract In this chapter, we show examples of multiplication with the method identified on Guamán Poma's yupana. In addition, we propose possible calculation methods for Inca addition, subtraction, and division, as deduced from the method of multiplication identified.

Keywords Yupana · Arithmetic operations

South American corn comes in a variety of colors (Fig. 8.8), supporting the use of color to represent the different values of ten when kernels of corn were used as counters in calculating.

Readers wishing to carry out the following examples should use counters of various colors on a checkered surface. Those who have the chance might use kernels of colored corn, just like the ancient Incas would have used. The direct manipulation of objects makes understanding much easier, and repeating the gestures of an ancient Quipucamayoc would certainly be more exciting.

Examples of Multiplication

As another example of the procedure, consider the multiplication: $21 \times 15 = 315$ (Fig. 11.1).

Assuming that color represents the powers of 10, we choose the colors at random and indicate the units with purple kernels, the tens with orange kernels, and the hundreds with blue kernels.

- The multiplicand, the number 21, is inserted into the rightmost column with one counter per square. This requires 3 squares (2 orange counters and 1 purple).
- While the multiplier, the number 15, can be subdivide in several ways, logic requires us to use the fewest columns in order to speed up the calculation, but at the same time, use the fewest counters per square to retain visual control and

C. Florio, *The Mathematical Enigma of Guamán Poma's Yupana*, https://doi.org/10.1007/978-3-031-92662-4_11

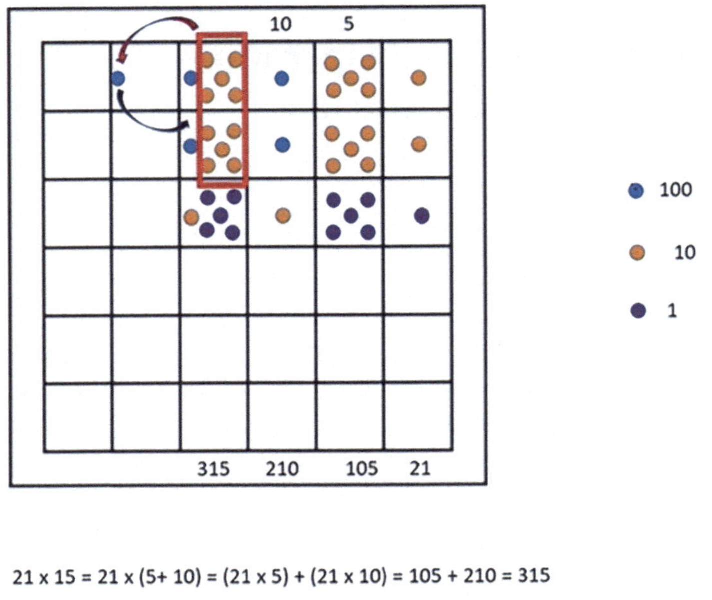

Fig. 11.1 Example of multiplication involving hundreds and three colors for the counters. (Image by the author)

decrease the likelihood of mistakes. We choose to subdivide 15 as 10 and 5. Ten counters in one square may seem like too many, but in reality, ten counters or kernels of one color can be replaced by a single kernel whose color means the next higher power of 10.

- After choosing 10 and 5 for the columns, into the column of 5 we insert the quintuple of the counters in the first column (10 orange and 5 purple), while into the column of 10 we insert the counters of the first column tenfold (2 blue and 1 orange). That is, ten times the first column yields 20 orange and 10 purple kernels, and these can be replaced and represented by 2 blue and 1 orange, as 10 orange correspond to 1 blue, and 10 purple correspond to 1 orange.
- In the last column, all the counters in the columns of 5 and 10 are collected (5 + 10 = 15, which is the multiplier) and counted (2 blue, 11 orange, and 5 purple). Any 10 counters of the same color are grouped and replaced by 1 whose color represents the next higher power of 10; this yields 3 blue, 1 orange, and 5 purple, which is read as the product, 315.

A final example is provided: 2,212 × 27 = 59,724 (Fig. 11.2).

- We use purple counters for the units, orange for the tens, dark blue for the hundreds, yellow for the thousands, and azure (light blue) for the tens of thousands.
- In the first (rightmost) column, using one counter per square, we insert the 7 counters that represent the multiplicand 2212: 2 yellow, 2 dark blue, 1 orange, and 2 purple. This requires seven rows.

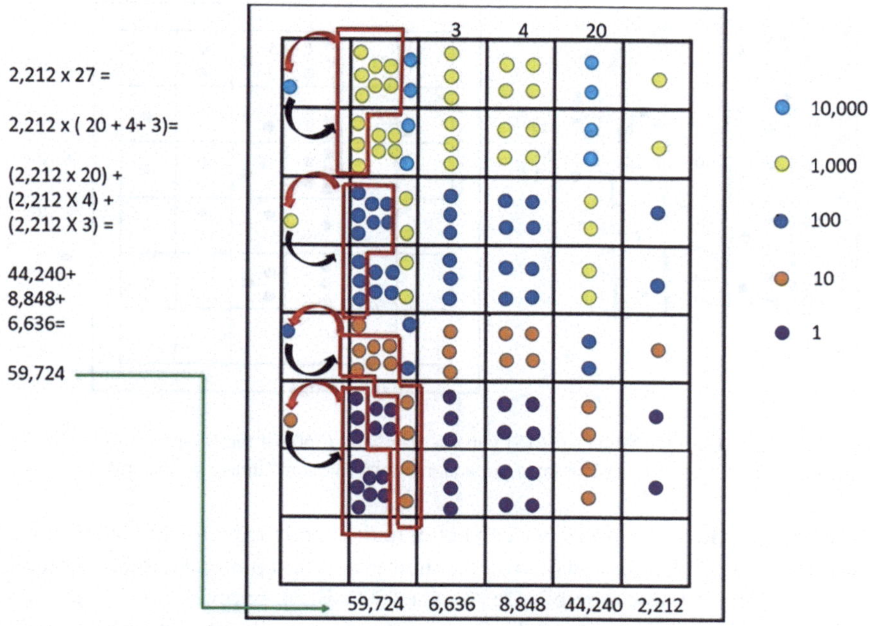

Fig. 11.2 Example of multiplication involving tens of thousands and five colors for the counters. (Image by the author)

- The multiplier, the number 27, is subdivided as 20 + 4 + 3. This requires 3 columns: one for 20, one for 4 and one for 3. Of course, different choices are possible (for example, subdividing 27 as 20 and 7 or 10, 10, and 7; however, remember the logic as previously described).
- Into the column of 20, we insert the 7 counters of the first column taken 20 times, starting with the first yellow counter at the top: Instead of inserting 20 yellow counters, we insert just 2 azure, as 10 yellow counters correspond to 1 azure. Continuing in this manner, 14 counters are inserted: 4 azure, 4 yellow, 2 dark blue, and 4 orange.
- In the column of the 4, we insert the counters of the first column taken 4 times.
- In the column of the 3, we insert the counters of the first column taken 3 times.
- Into the last column, all the counters of the columns of the 20, 4, and 3 are gathered. We replace any ten counters of one color with a single counter representing the next higher power of 10. In this particular multiplication, there are four substitutions: ten purple counters are replaced by one orange, ten orange by one dark blue, ten dark blue by one yellow, and ten yellow by one azure.
- The product is 5 azure counters, 9 yellow, 7 dark blue, 2 orange, and 4 purple. These colors represent the number 59,724.

In comparison, our Hindu–Arabic numerals appear bland and arid when compared to this chromatic picture. Purely for pleasure, Guamán Poma's yupana is represented with colored kernels in Fig. 11.3:

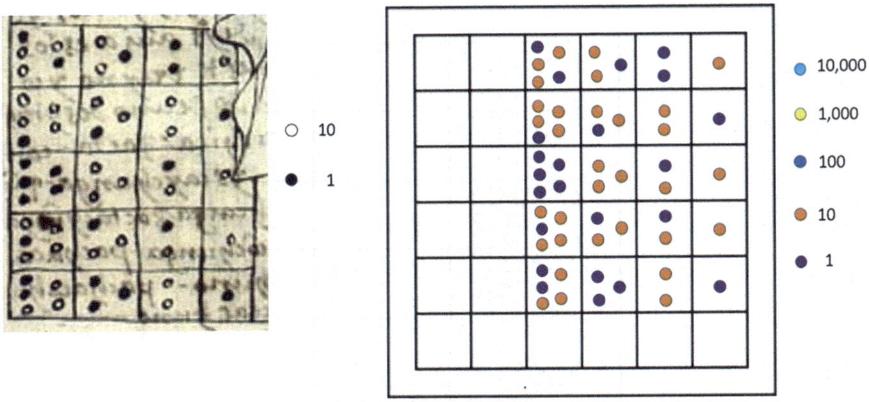

Fig. 11.3 (Left) Guamán Poma's (1615) yupana, page 362 (360) of the *Nueva Corónica*. (Image in the public domain). (Right) The same yupana, redrawn in color. (Image by the author)

Looking at these examples, we can better understand the words of Father Acosta when he wrote, "[The Indians] will take their grains and put one here, three there, and eight I don't know where" (De Acosta 1590, Book 6, pp. 411–412, as translated in Martin and Wasserman 2010, pp. 5–6). Above all, the word "exchanged" ("trocarán") might refer to the replacement of ten kernels with one of different color indicating a higher power of 10. One can also understand, looking at these colorful provisions, how chroniclers who were not expert calculators might have been confused and fascinated by a functional calculation so different from the linear counting boards and written numbers prevalent in Europe at the time (Pullan 1968).

It is possible that simple addition and subtraction tasks involving numbers represented on a quipu were performed mentally. However, it is likely that such tasks were conducted with kernels, perhaps even without formal organization on a yupana, in a manner suggested by the operations shown in Guamán Poma's drawing. These are discussed in turn, using purple counters valued at 1, orange valued at 10, blue valued at 100, and yellow valued at 1000.

Addition

For the operation of addition (A1), the kernels representing the addends are combined, simplified, and read (Fig. 11.4).

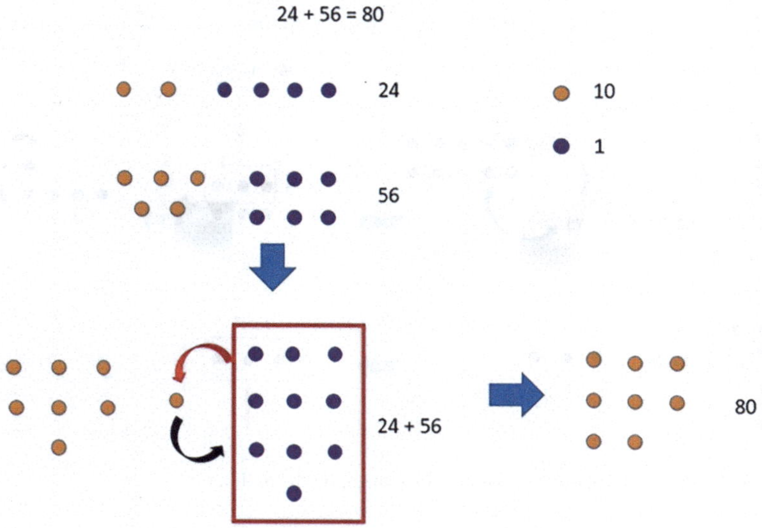

Fig. 11.4 Steps in addition (Example A1). (Image by the author)

Example A1: 24 + 56 = 80 (Fig. 11.4)

Steps

1. Place the counters representing the addends 24 and 56 on a flat surface.
2. Combine all the counters.
3. Exchange any 10 counters of the same color (if any) for a single counter representing the next higher power of 10.
4. Read the sum.

Subtraction

For subtraction, we will consider two alternative algorithms (S1 and S2).

First Alternative Subtraction Method (S1) An equal number of counters are removed from both the minuend and subtrahend at the same time (Fig. 11.5). This method eliminates the need to keep track of the subtrahend mentally, but also involves laying out and moving additional kernels, relative to the alternative discussed in S2 below.

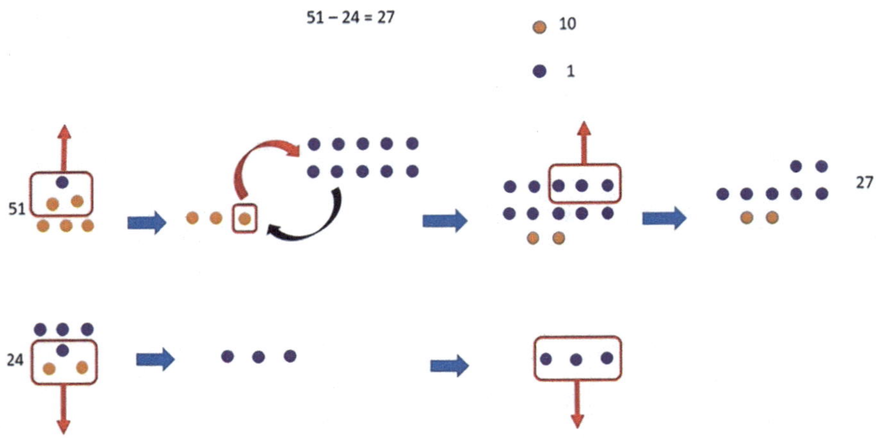

Fig. 11.5 Steps in subtraction (Example S1). (Image by the author)

Example S1: 51 – 24 = 27 (Fig. 11.5)

Steps

1. Align the counters representing the minuend and subtrahend (in this case, 51 and 24) on two different levels.
2. Eliminate an equal number of counters from both concurrently. In the example, this would remove two orange counters and one purple from both numbers. The aim is to make the subtrahend (24) disappear completely.
3. Three purple counters remain in the subtrahend. To eliminate them, three purple counters must be removed from both groups. However, the minuend at this point consists of three orange counters. To permit further removal, exchange 1 orange counter in the minuend with 10 purple counters.
4. Remove three purple counters from each group. This fully removes the subtrahend.
5. As the subtrahend has now fully disappeared, read what remains of the minuend: 2 orange counters and 7 purple, or 27. Thus, the difference is 27.

Second Alternative Subtraction Method (S2) Remove the subtrahend from the minuend (Fig. 11.6). While this method involves fewer kernels and movements, relative to the method in S1, it has a higher potential for error, as the number of counters to be eliminated must be kept track of mentally; the potential for error increases with the size of the numbers and number of exchanges involved.

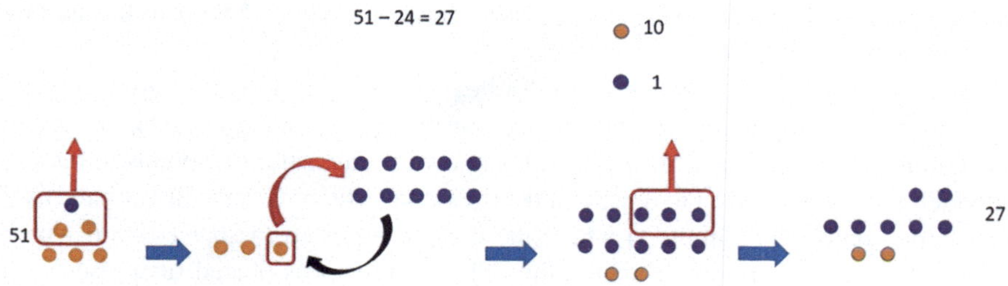

Fig. 11.6 Steps in subtraction (Example S2). (Image by the author)

Example S2: 51 – 24 = 27 (Fig. 11.6)

Steps

1. Align the counters of the minuend on a flat surface: 51, or 5 orange and 1 purple counters.
2. The subtrahend—the number 24, represented by 2 orange and 4 purple counters—must be removed from the minuend. Start by removing two orange counters, which represent all that are needed for the subtraction, and then the one purple counter, as there are currently no others. This leaves another three purple counters to be removed.
3. At this point, since only orange counters remain in the minuend, one must be exchanged for ten purple counters.
4. Once the exchange has been made, remove the final three purple counters.
5. Read the difference in the remaining counters: 2 orange and 7 purple counters, or 27.

We could also optimize this method by retaining the removed counters and keeping them in view, in order to be able to control the value as it was removed and then subtracted, but we would still need a way of not forgetting and keeping in mind the value to be subtracted.

Division

For division, the chronicler Garcilaso de La Vega suggests that the Incas knew how to divide: "the calculations were so exact that I hardly know whether to praise the more the accountants who made their reckonings without the use of figures and contrived to divide exactly very small quantities, a thing our mathematicians have great difficulty in doing, or the royal governor and officials who followed the process with perfect ease"[1] (De la Vega 1609, Book V, Chapter XIV, as translated by Livermore 1966, p. 275).

Since division is the inverse of multiplication, we can derive an algorithm by reversing the multiplication found on Guamán Poma's yupana, left to right instead of right to left. In this case, the dividend is known because it corresponds to the old product; the counters used to represent the dividend (which we have in our hand) are distributed onto the columns in which the divisor is placed (as parsed) and which correspond to the columns in which the old multiplier was placed (as parsed). To finish, the column of 1 is constructed, which represents the quotient and corresponds to the old multiplicand. The only difference is that the dividend counters must be inserted into the columns of the divisor using three rules:

1. Proceed horizontally, rather than vertically.
2. Insert the counters of higher value first into the columns representing the highest power of 10.
3. The counters of the same row, but of different columns, must have colors consistent with the multiples of 10, as reported on the columns themselves.

Here are several examples, in order of increasing difficulty (D1–D4):

Example D1: 360 ÷ 12 = 30, with 0 Remaining (Figs. 11.7, 11.8 and 11.9)

Steps

1. Collect the counters that form the dividend, the number 360: 3 blue and 6 orange counters.
2. Subdivide the divisor, the number 12, as 10 and 2 (10 + 2 = 12). This produces a column of 10 and a column of 2.
3. Insert the counters of the dividend *horizontally* into the divisor columns, using the highest values first in each case. The first dividend counter to be inserted is the one with the highest value, blue, and it is inserted into the divisor column with the highest value, which is 10.
4. Continuing horizontally, proceed to the divisor column of 2. Here two counters must be inserted, but they must represent the next lower power of 10 than the previous (leftmost) column. This yields the following layout (Fig. 11.7):
5. With the same procedure, fill in the next two rows using the remaining counters (Fig. 11.8).

[1] "los contadores, q' sin cífras de guarismos haziá sus cuentas y partíciones tá ajustadas de cosas tan menudas, cosaq' nuestros Arísmetícos suelé hazer có mucha dífficultad: o al Gouernador y mínístros regíos, q' cen tanta facílidad entédian la cuenta y razon, que de todas ellas le dauan" (De la Vega, 1609, Book V, Chapter XVI).

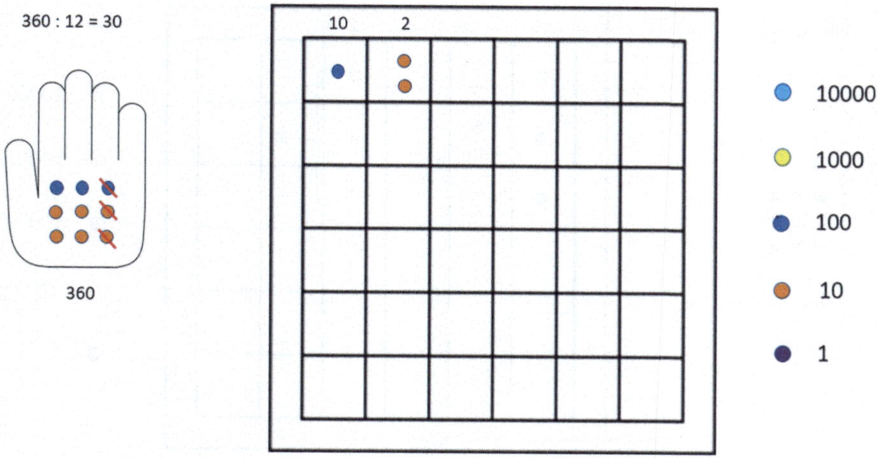

Fig. 11.7 Initial yupana layout in division (Example D1). (Image by the author)

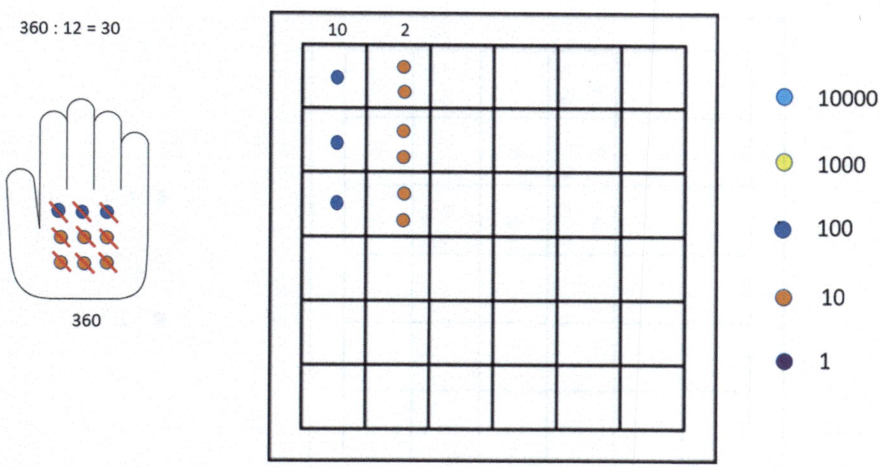

Fig. 11.8 Second yupana layout in division (Example D1). (Image by the author)

6. As all the counters representing the dividend have now been placed in the divisor columns, the column of 1 (rightmost) must be constructed, which will display the quotient. New counters are selected to build it from a separate container. To build the column of 1, take as a reference one of the columns already built on the yupana; it is best to take the one with the lowest value. In this case, this will be the column of 2. In the topmost square of the column of 2, there are two orange counters, so in the topmost square of the column of 1, one orange counter must be inserted, and so on for the other two rows. This will yield three orange counters in the column of 1 (Fig. 11.9), or the number 30, which is the quotient.

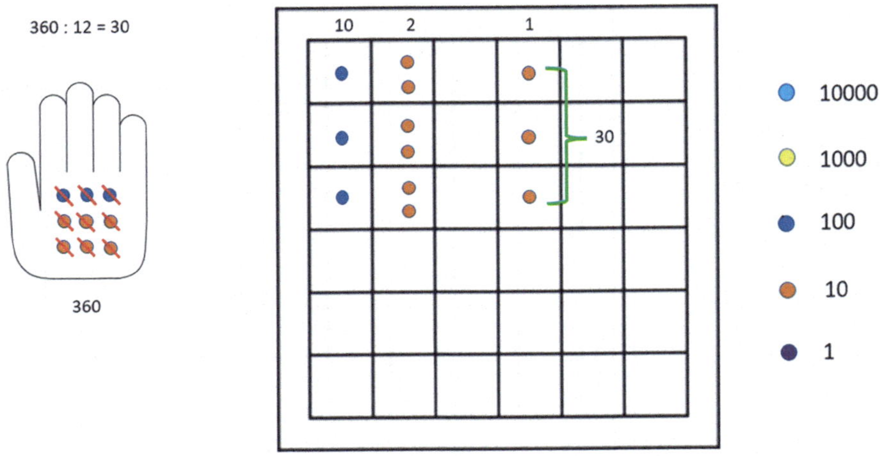

Fig. 11.9 Final yupana layout in division (Example D1). (Image by the author)

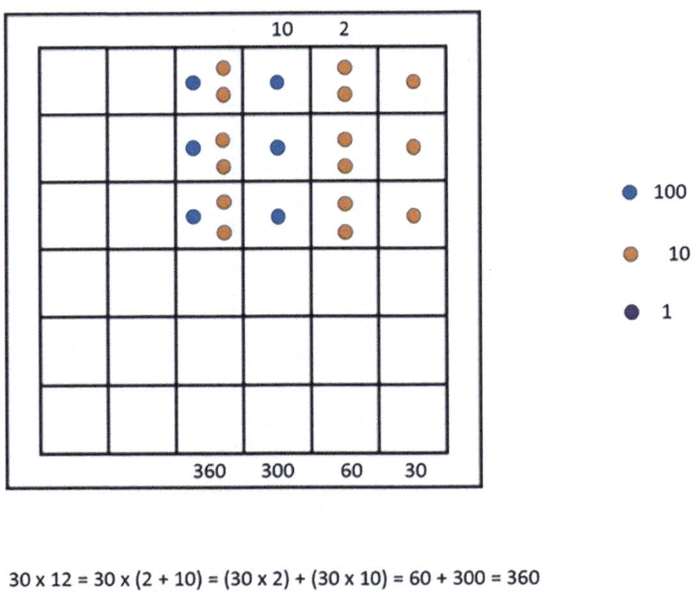

Fig. 11.10 Multiplication as inverse of Example D1 division. (Image by the author)

That this is the inverse of multiplying $30 \times 12 = 360$ can easily be verified by carrying out the multiplication on the yupana (Fig. 11.10) and comparing it to Fig. 11.9. The only difference is that the kernels that form the number 360 are in the leftmost column, rather than the hand.

Of course, we cannot be sure that division was carried out in this way, but it is very likely, since the reverse process between multiplication and division is found in many ancient civilizations that used an additive number system. We can also note that if the division had actually been carried out in this way, the arrangement of the counters on the yupana would necessarily have had to respect a more rigorous order than that of the multiplication carried out by Guaman Poma.

Example D2: 364 ÷ 12 = 30, with 4 Remaining (Fig. 11.11)

Steps

1. Collect 3 blue, 6 orange, and 4 purple counters to represent the number 364, the dividend.
2. Subdivide the divisor, the number 12, into 10 and 2 just as was done previously in Example D1, obtaining again the columns of 10 and 2.
3. Insert the 3 blue and 6 orange counters representing the dividend into the divisor columns of 10 and 2, just as was done previously in Example D1.
4. In the hand, four purple counters remain. We might insert one into the column of 10, as this is the counter of the highest value available, but in this case, there would not be corresponding counters of lower powers of 10 to fill in the column of 2. The purple counter represents the unit, so there are no lower powers of 10. (Obviously, from a mathematical perspective, it would be possible to continue by selecting colors to represent tenths, hundredths, thousandths, and so on, and continuing the operation by exchanging counters; however, there is no historical or archaeological evidence of the Incas using fractions.) As the four purple counters cannot be inserted onto the yupana, they remain in the hand as the remainder.
5. Construct the column of 1 and read the quotient (Fig. 11.11): the number 30, with a remainder of 4.

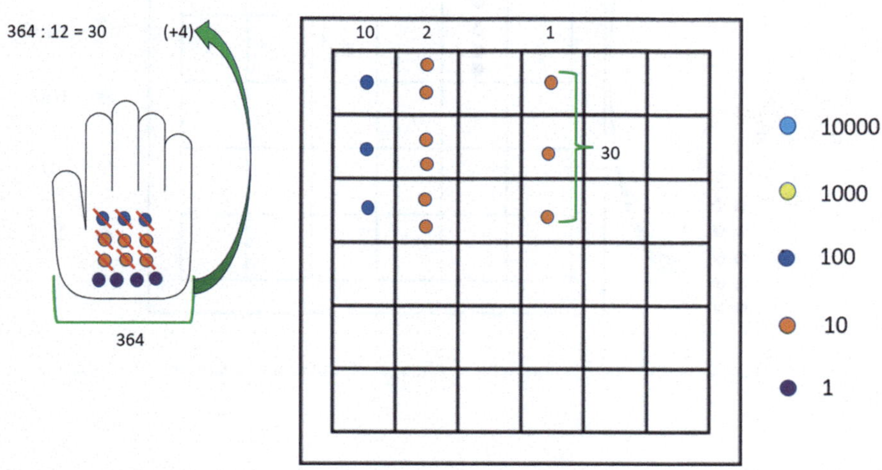

Fig. 11.11 Yupana layout in division (Example D2). (Image by the author)

Example D3: 164 ÷ 13 = 12, with 8 Remaining (Fig. 11.12)

Steps

This example requires the exchange of counters between powers of 10.

1. Collect the counters that represent the dividend, the number 164: 1 blue, 6 orange, and 4 purple counters.
2. Subdivide the divisor, the number 13, as 10 and 3 to produce columns of 10 and 3.
3. In the first row of the column of 10, enter the counter with the highest value, which in this case is blue. Continuing horizontally, insert 3 orange counters into the column of 3 (the next lower power of 10).
4. Fill in the second row. As the counter with the highest value is now orange, insert it into the column of 10. Into the column of 3, insert 3 purple counters.
5. Fill in the third row by inserting one orange counter. However, to fill in the column of 3, only 1 purple counter is available, rather than the 3 needed to fill the square. There is still 1 orange counter left; this is changed to 10 purple, so the 2 counters needed are now available. This leaves eight purple counters, which represent the remainder.
6. Build the column of 1 taking as a reference the column of 3. This produces 1 orange and 2 purple counter, the number 12.
7. The quotient is 12 with a remainder of 8.

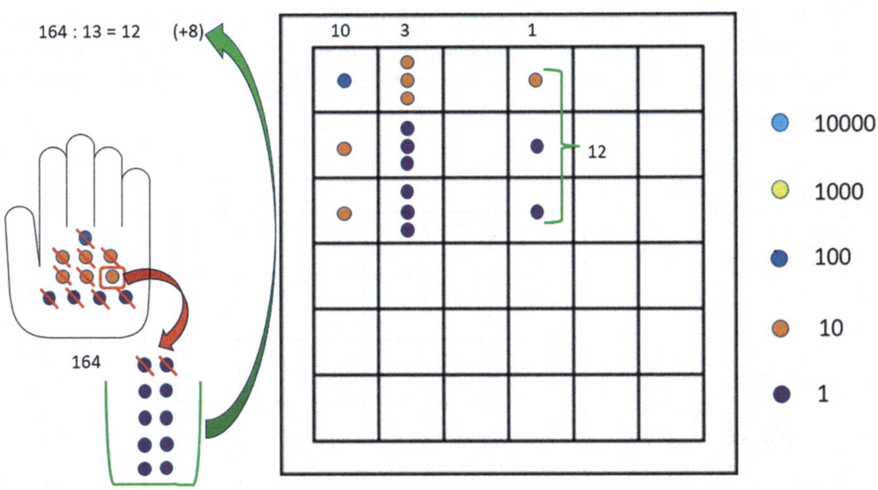

Fig. 11.12 Yupana layout in division with exchange of counters (Example D3). (Image by the author)

Example D4: 1,025 ÷ 213 = 4, with 173 Remaining (Fig. 11.13)

Steps

This example requires an initial exchange of counters.

1. Collect 1 yellow, 2 orange, and 5 purple counters to represent the dividend, the number 1025.
2. Subdivide the divisor, the number 213, into 200, 10, and 3. This produces columns of 200, 10 and 3.
3. Into the column of 200, we must insert 2 counters of the same color with the highest value. However, in this case, there is only one counter with the highest value, the yellow one valued at 1000. Exchange it for ten blue counters.
4. Insert 2 blue counters into the column of 200. Continuing horizontally, insert 1 orange counter into the column of 10 and 3 purple into the column of 3.
5. Fill in the second row using the same procedure. However, the purple counter needed to complete the sequence is not available. In the hand, exchange 1 blue counter for 10 orange; then exchange 1 orange with 10 purple in the hand to complete the second row.
6. Fill in the remaining rows using the same procedure (exchanges are not required).
7. At this point, only one counter of higher value remains, a blue counter. If this were exchanged for orange counters, the column of 3 could not be completed because there would not be counters with color below the units. The inability to proceed further means that the counters left represent the remainder: 1 blue, 7 orange, and 3 purple, the number 173.

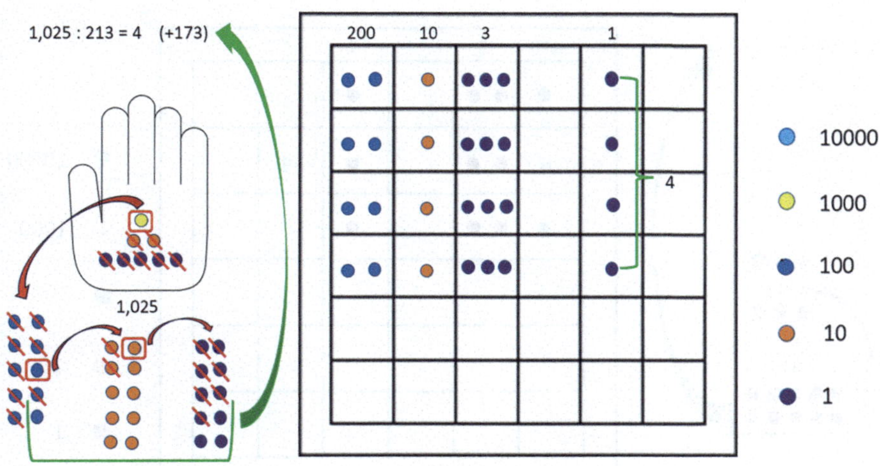

Fig. 11.13 Yupana layout in division with initial exchange of counters (Example D4). (Image by the author)

8. Build the column of 1, taking as a reference the column of 3; this yields 4 purple counters, or the number 4.
9. The quotient is 4 with a remainder of 173.

Example D5: 313 ÷ 102 = 4, with 7 Remaining (Fig. 11.14)

This example uses a divisor that when broken down contains non-sequential powers of 10, such as hundreds and ones (there are no tens).

1. Collect 3 blue, 1 orange, and 3 purple counters to represent the dividend, the number 313.
2. Subdivide the divisor, the number 102, into 100 and 2. This produces columns of 100 and 2.
3. In the first row of the column of 100, enter the counter with the highest value, which in this case is blue. Continuing horizontally, insert 2 purple counters into the column of 2, because there is a jump of 2 powers of 10 from 100 to 2, and therefore two colors are skipped on the scale.
4. In the second row, insert a blue counter into the column of 100. Continuing horizontally, you should insert two purple counters, but we only have one. We insert it. To get the second, we exchange the orange counter with 10 purple counters from which we take the necessary counters and insert it into the second row in the column of 2.
5. In the third row we insert a blue counter in the column of 100 and 2 purple counters in the column of 2.

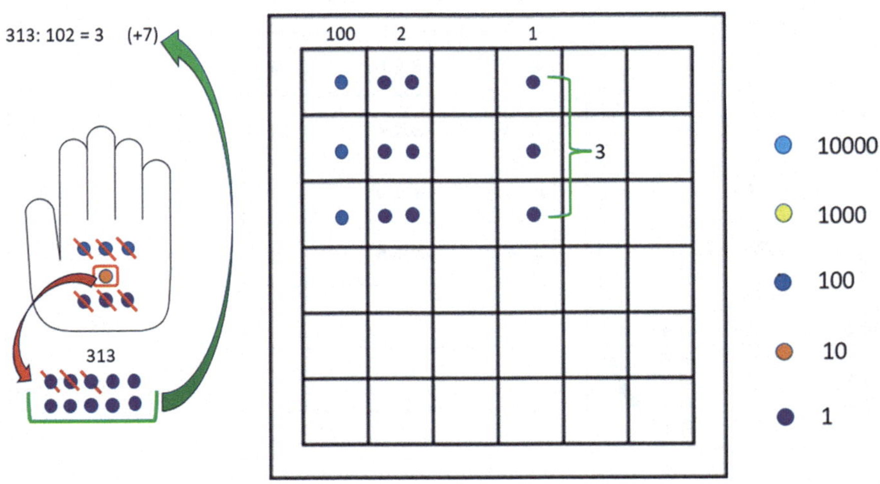

Fig. 11.14 Yupana layout in division with non-sequential powers of 10 (Example D5). (Image by the author)

6. There are only 7 purple counters left in the hand that cannot be inserted into the yupana, because a purple counter would go into the column of 100, but there would be no colors below purple to fill the column of 2. So the seven purple counters represent the remainder, or the number 7.
7. Build the column of 1, taking as a reference the column of 2; this yields 3 purple counters, or the number 3.
8. The quotient is 3 with a remainder of 7.

Competing Interests The author has no conflicts of interest to declare that are relevant to the content of this chapter.

References

De Acosta, J. (1590). *Historia natural moral de las Indias, en que se tratan las cosas notables del cielo, y elementos, mentales, plantas, y animales dellas: y los ritos, y ceremonias, leyes, y gouierno, y guerras de los Indios*. Juan de Leon.

De la Vega, G. (1609). *Commentarios reales, que tratan del origen de los Yncas, reyes que fueron del Peru, De su idolatria, leyes, y gouierno en paz y en guerra: de sus vidas y conquistas, y detodo lo que sue aquel Imperio y su Republica, antes que los Españoles passaran a el*. Pedro Crasbeeck.

Livermore, H.V. ; De la Vega, G. (1966). *Royal commentaries of the Incas, and general history of Peru. Part one* (H. V. Livermore (trans.)). University of Texas Press.

Guamán Poma, F. de A. (1615). *El primer nueva corónica y buen gobierno*. Museum Tusculanum Press.

Martin, C. E., & Wasserman, M. (2010). Writing and record-keeping in the ancient Americas: The quipus of the Andes. In *Latin America and its people: Volume 1, To 1830* (2nd ed., pp. 1–13). Prentice-Hall.

Pullan, J. M. (1968). *The history of the abacus*. Hutchinson & Co.

Appendix A: Probability of Randomly Inserted Counters

Abstract This appendix explains the probability of randomly inserting black-and-white counters into the first three columns (starting from the right) of Guamán Poma's yupana to form a number, its double, and its triple.

Keyword Probability

The rightmost column of the yupana, also known as the column of 1, contains a single counter in each box (hence, column of 1). With a single counter and two colors, it is possible to produce six numbers (Table A.1):

The corresponding doubles and triples are given in Table A.2. These would fill in the second and third columns on the yupana, the columns of 2 and 3 (note that on the yupana, these columns are read from right to left, while the table below is read left to right).

The probability (Pt) that the three columns would contain a number, its double, and its triple is calculated by summing the probabilities for the six cases (Table A.2, a–f)

$$\mathrm{Pt} = \mathrm{Pa} + \mathrm{Pb} + \mathrm{Pc} + \mathrm{Pd} + \mathrm{Pe} + \mathrm{Pf}$$

Table A.1 Six combinations of black-and-white counters, where white counters are valued at 10 each, black counters as 1

Black	White	Number
5	0	5
4	1	14
3	2	23
2	3	32
1	4	41
0	5	50

C. Florio, *The Mathematical Enigma of Guamán Poma's Yupana*,
https://doi.org/10.1007/978-3-031-92662-4

Table A.2 Singles and corresponding doubles and triples

Single	Double	Triple	Case
5	10	15	a
14	28	42	b
23	46	69	c
32	64	96	d
41	82	123	e
50	100	150	f

The probability of each case is the product of the probabilities of every single number of that term being represented, namely:

$$Px = Px_1 \times Px_2 \times Px_3$$

where x represents the values in Cases a–f.

Since (see Table A.1)

$$\text{Pa} = \text{Pf},$$

$$\text{Pb} = \text{Pe},$$

$$\text{Pc} = \text{Pd},$$

and

$$\text{Pa} = \frac{\binom{5}{5}}{2^5} \times \frac{\binom{10}{10}}{2^{10}} \times \frac{\binom{15}{15}}{2^{15}} = 9.3132257 \times 10^{-10}$$

$$\text{Pb} = \frac{\binom{5}{4}}{2^5} \times \frac{\binom{10}{8}}{2^{10}} \times \frac{\binom{15}{12}}{2^{15}} = 0.000095344$$

$$\text{Pc} = \frac{\binom{5}{3}}{2^5} \times \frac{\binom{10}{6}}{2^{10}} \times \frac{\binom{15}{9}}{2^{15}} = 0.009788665$$

The total probability is therefore:

$$\mathbf{Pt} = 2\text{Pa} + 2\text{Pb} + 2\text{Pc} = \left(2 \times 9.3132257 \times 10^{-10}\right) + \left(2 \times 0.000095344\right) + \left(2 \times 0.009788605\right) = 0.0198 = 20/1{,}000 = \mathbf{2/100}$$

We can simplify the calculation by considering the rightmost column to represent the number 32 (that is, we exclude the possibility that other numbers were represented). Given the number 32 in the first column, we then calculate the probability that the two adjacent columns will contain its double and triple (i.e., 64 and 96). The calculation is as follows:

$$\mathbf{P} = \frac{\binom{10}{4}}{2^{10}} \times \frac{\binom{15}{6}}{2^{15}} = 0.03 = \mathbf{3/100}$$

and the result is **3/100.**

The manufacturer's authorised representative in the EU is Springer Nature Customer Service Centre GmbH, Europaplatz 3, 69115 Heidelberg, Germany. If you have any concerns regarding our products, please contact ProductSafety@springernature.com

Printed and bound by CPI Group (UK) Ltd, Croydon, CR0 4YY
30/09/2026
02225577-0002